The Future Is Imagined

Russell John Connor

Published April 2013 by emp3books Ltd
Norwood House, Elvetham Road, Fleet, GU51 4HL

Printed and Bound by Lightning Source.

Set in Bookman Old Style: Cover Design; Linda Eglite

Printed on acid-free paper from managed forests. This book is printed on demand, so no copies will be remaindered or pulped.

ISBN 978-1-907140-73-0

Dedicated to Peteris Vasks

Through his music it is clear that there is
more to life than buying and selling.

'Imagination is more important than knowledge. For knowledge is limited to all we now know and understand, while imagination embraces the entire world, and all there ever will be to know and understand.'
Albert Einstein

Contents

Synopsis

The Future Is Imagined is a personal journey through a set of ideas drawn from philosophy, economics, mathematics, cognitive science and social science. These bring the slippery characters of Time, Complexity and Uncertainty centre-stage and enable the whole person and the essence of *being* to be fully appreciated.

These powerful ideas shine a light on a blind spot in social and management science. Everybody thinks they know about experience. In fact, we don't really. That doesn't stop us though from forming theories about how we do things and the thinking process that brings this about. It is a short hop from there to harbour deep-seated but misguided views on human nature.

What is lacking, and not just in management science, is a unifying idea about human nature. What is certain is that our ideas about human nature at work are clouded by serious, invasive misconceptions that have continually fuelled the development of ill-advised economic and managerial systems. It is clear that these shortcomings flow from theories and models that do not embrace the whole person.

The powerful ideas that are explained in simple terms are used to construct a new model of human capability. This provides the platform for a radical reappraisal of virtually all that currently stands for Human Resource Management and it has implications for evolutionary biologists, economists, psychotherapists, politicians and anyone interested in how to build a healthy society.

This is not a book that brings religion in by the back door, nor is it about being person-centred with a plea to give individuals unwavering and unconditional positive regard. It takes into account the darker side of humanity and yet argues that, in treating people as people, the total adds up to being greater than the sum of the parts.

1

Foreword

As a business manager, I have in the past received good advice from a range of people. The very best has been about focus. Know your audience; understand what they need. My advisors have all challenged me to think about whom the products and services are for? Taking their advice I have used the same thinking to narrow the audience down for this book. My advisors will be pleased to hear that it is focused on.... everyone!

Everyone that is who's interested in life and living organisms. This includes biologists, economists, managers, psychotherapists, human resource professionals, those responsible for building organizations and policy-makers who affect how we all live our lives.

So what makes me arrogant enough to think that I can appeal to such a wide audience? Well, this book is about some universal things. It is about freedom and choice and how we handle uncertainty. It addresses how we think about thinking and how our thinking has provided systems that are based on erroneous views. This book tackles head-on the vital subject of human nature.

If that is generally what it is about, what kind of book is it?

This is not an academic textbook and it is not written specifically for academics. It does though provide a broad range of social science and research-based ideas.

It's not a textbook but it doesn't go to the other extreme and provide the ten steps to being a better entrepreneur/artist/member of the human race. It is not a cookbook with recipes for instant results.

The book contains plenty of my life experiences so is it an autobiography? No. I add in my reflections because it is important to know how the theory that I expound has developed and the different strands that have contributed to this.

So, it's not an autobiography, a cookbook or an academic textbook but it might be the longest resignation letter yet from the world of work! At the time of writing, I am a management consultant plying my trade in helping leaders simplify people and organizational management. Despite investing heavily in software development, business, you might say, is not exactly booming. I put this down to providing a service based on a different premise to that offered by virtually all other consultancies. In the busy world that CEOs and HR Directors inhabit I have found it nearly impossible for them to slow down enough to take on board that there is a different way. As such they stick with the familiar and overlook the reality that if you keep doing what you always do, you get what you always got.

After you read the book, you might say that my business is not booming because, 'Russell, old bean, you have lost touch with reality'.

My hope is that by reading the book, you will have given yourself the time to imagine a very different future.

Acknowledgements

I have the great privilege of being able to call one of the world's greatest living composers a friend. Peteris Vasks is composing some of the most sublimely beautiful pieces ever in the history of classical and contemporary music. I am not just name-dropping, Peteris and his music is an inspiration. Through having a glimpse of how he crafts his work and nurtures it and finally lets it out into the world, I realise the true genius of the man and the dedicated effort that goes with it.

In recognizing the hard work that Peteris puts into his compositions and the joy it brings to him and his appreciative audiences, Peteris has kept me going on my own creative path longer than I otherwise might. That is not to compare what I am producing with that of Peteris. I am more of an arranger than a true composer. In street terms I sample and dub. So whose work am I sampling? The work is from a range of people; some famous and some not famous enough. I have named them all in this book and have provided some context to describe when I was first introduced to the ideas and what was happening at the time. Nevertheless, I have tried to ensure that the collection of ideas comes over with a consistent theme rather than simply a collection of 'greatest hits'. Indeed, I do go on to create a little of my own music. When the discordant notes cut in you will probably recognise my work!

The main figure whose work reverberates around my book is Elliott Jaques. Elliott who? I hear you ask. You obviously haven't read my book, *It's About Time.* In this I have written extensively on Elliott, his ideas and the implications of his work. Here, I try to avoid going over the same ground although I repeat a few key principles. Whilst there are recurrent themes *The Future is Imagined* is not the same material orchestrated in modified form.

One of the key themes of the book is about choice. We are free to choose. *What* we choose is however shaped by a variety of factors. I have chosen to write this book but its content is shaped by my upbringing in a period of time characterised by high levels of trust. Politicians and policy-makers still held vestigial beliefs that it was possible to build a kingdom fit for heroes. It has been shaped through an educational system that valued creativity, inquiry and analytical analysis. My friends and supporters have shaped it. It has also been shaped by my past bosses and colleagues.

So this is a big thank you to all those in my life because, in their own way, they have helped me compose this book.

Russell Connor
London, 2013

4

Introduction

You're invited on a journey. Some of it is retracing a path that I have already taken and some of it involves standing on the shoulders of giants to look at the route ahead.

Before we start the journey hand in hand, I am sure that you want to know whether we are going on a nice ramble with the opportunity to look at many different views and smell the roses on the way. Sorry, not quite! This is difficult terrain through human nature.

Of course, before you begin you will want to know with whom you will be travelling. Well, I am a middle-aged, white, British male and let's not mention the hairline! In terms of career, I have spent most of my working life in human resource management. No, please don't stop here, I have some redeeming qualities! I am fortunate to have had a good education and I have worked in various countries and very different types of organizations. Just like a traveler returning from unexplored parts I have brought back some amazing findings, which I want to share with you. Don't worry though these treasures are more philosophical gems than directly connected with HR.

On our journey there will be plenty of undergrowth to chop through and the main view, although magnificent, will only be glimpsed. What is this view that is worth such effort? The scene that we will be viewing from a number of angles is the Twin Peaks of Freedom and Choice.

Let me explain with a story. When the Lord (no, not white British) created the world and the plants and animals to live in it, a process that took a very long time, He faced a certain dilemma, which He voiced as; 'To be predictable and orderly or unpredictable and chaotic? I could make everything nice and neat, in other words, predictable. However, there is a problem with this I see. These humans, whom I have made pretty smart, will undoubtedly learn to predict everything. When that happens they'll have no motive to do anything at all because they will recognise that the future is totally determined and cannot be influenced by their action.'

The Lord carried on his musing; 'Creating such order won't work but, on the other hand, I could make the world unpredictable.' Then again He saw the catch; 'If I make everything unpredictable they will gradually discover that there is no rational basis for any decision whatsoever. Thereupon they again will have no motive to do anything at all. Humans; what have I created!'

Cutting through the dilemma, He realised that neither scheme would make sense. He concluded; 'I must therefore make a mixture of the two. Let some things be orderly and predictable therefore outside of the human capacity to alter and let some things unpredictable and open to the influence of free will. These humans will then, amongst other things, have the very important task of finding out which is which.'

Finding out which is which, it turns out, is what my life's journey is about and I think that it's the life journey of many others.

Finding out which things we can control and which we cannot is about discovering what space we really can take up, how much influence we have and at what point are we carried along like a reed in a fast flowing stream.

On our journey together we will be looking from different angles at free will and the limits of choice. Determining the limits of free choice is a very important task particularly today when people are trying to devise machines, full of artificial intelligence, to foretell the future.

We will meet slippery characters travelling under the name of Time, Complexity and Uncertainty. They are so slippery that when you're first introduced to them they seem like the stuff of physics. But as we look through the lens of human experience we begin to see these characters for what they are and how they deeply influence our lives.

In Part One, you are invited to retrace the journey that I have been on as I introduce eminent psychologists, philosophers, mathematicians, economists and social scientists; all of whom have a great deal to say about *time, complexity, uncertainty* and the nature of *being*. We will be taking a wide path and sometimes diverting off this for brief periods. If it feels like at any time we have disappeared down a rabbit's hole, please bear with me, it all comes together at the end. I will expand on the key themes at various stages of my life, and to set the context, I will provide details on the key struggles that were relevant at the time. These examples illuminate the fact that life carries on between two polarities but more of this in a short while.

In Part Two, I will provide a new model of human capability that is drawn from the influential ideas I have gathered en route. This model also reveals some of the stories that we have told ourselves, which have helped prevent a full understanding of our thinking capability and of human nature.

From a vantage point made accessible by sitting on the shoulders of intellectual giants we will, in Part Three, look at the route ahead. We will use the ideas that I have introduced to help us question received wisdom and think about what the world of work, indeed society, would be like if our work systems, processes and our attitudes were built on the in-depth appreciation of the individual; of the person as a real human being; living in a complex world and capable of making decisions without full information - but not wholly rational ones.

Part One
The Journey

Chapter 1
In The Beginning

In 1975, therefore a long time ago, I arrived at University College London for a selection interview on to their undergraduate psychology course. As I pushed the door open of a basement office in Gordon Square, I wondered whether I was the subject of an experiment rather than a candidate for an interview. The door stopped abruptly against a filing cabinet seemingly to prevent further access. By forcing myself against the cabinet and shutting the door behind me I managed to squeeze into the room and then my suspicions about an experiment were heightened. Sitting behind a desk were three strange looking men. One had a patch over the eye, another was missing an arm and the third was hiding behind a scruffy beard.

It turned out that it was a genuine interview and I had just met one of the people that would become a personal tutor, Professor Norman Dixon (the one with the missing arm). Also interviewing me were Professor Jonckheere (the one with the patch) and Dr Rawlins (the scruffy beard). It turned out that Professor Dixon was extra-ordinarily interesting and had written one of the best management books ever, *The Psychology of Military Incompetence*. It should be mandatory for all business schools. Dr Jonckheere went on to question the validity of the statistics on intelligence tests that underpinned the wide application of the Eleven-plus examination in the UK. Dr Rawlins introduced me to the 'delights' of statistics and the fascinating idea that the subject was not about studying what happened but establishing what else might have happened.

I put my passing the interview down to a conversation that we had about free will versus determinism. I'm not sure how we got on to the subject but it may well be because I did a bit of name dropping and mentioned one of the fathers of modern psychology, William James.

'Human beings, by changing the inner attitudes of their minds, can change the outer aspects of their lives'. William James

At one period in his life William James became acutely aware of the contingencies of life and the impact of chance. It occurred to him that we might not be in control of our lives. Troubled by this, James searched for an answer and it was an essay by a French Philosopher, Renouvier that grabbed his attention. After reading Renouvier's definition of free will, 'the sustaining of thought because I choose to when I might have other thoughts', James determined that free will was not an illusion and wrote, 'my first act of free will shall be to believe in free will'.

If truth were told, I was interested in a slightly different question to that which James posed to himself; one that had real life and death implications. At the time, I believed in the power of prayer and wanted desperately for God to intervene and cure my father of cancer. As well as the more academic interest in free will versus determinism, I was interested in the practice-based research regarding the extent of divine intervention.

In their own way the three professors that I met at the interview were also tackling similar questions (admittedly not divine intervention) including how much of our behaviour is determined by genes and how much is influenced by the environment. I didn't realise it at the time but addressing the genetic constituency of behaviour was fraught with danger for them. The word 'paradigm' hadn't appeared in my lexicon when I arrived at university but it is clear now that we all work within a framework, a way of seeing the world and acting within this. That is certainly true of academia. In the 1950s psychology was dominated by behaviourism, the school of thought popularized by John Watson and B. F. Skinner. Mental terms like 'know' and 'think' were branded as unscientific; 'mind' and 'innate' were dirty words. Behaviour was explained by a few laws of stimulus response learning that could be studied with rats pressing bars and dogs salivating to tones. Even voicing the idea that human capability was genetically determined was to risk being cast out into the academic wilderness. This paradigm still cast a long shadow when I entered the Psychology Department at University College London.

Whilst paradigms are useful and relevant, it is important to consider what the boundaries preclude and how these blinker a person so that they cannot consider any other perspective. In Chapter 9 we will look in detail at prevailing paradigms and their limiting beliefs.

The Big Ideas

I arrived at university with an 'A level' in biology, which meant that I had a basic grounding in the theory of evolution and its genetic mechanisms. It was into these topics that I gladly immersed myself through, what was termed at the time, Psychobiology.

Evolution and genetics together are potentially the big explainers. Darwin's big idea is essentially quite simple, evolution *is* non-random natural selection. We see evolution all the time when it is non-random selection through the human hand as in the cases of dogs or varieties of flowers. Yet, I had all the classical anxieties when it came to accepting that a similar process was at work when organisms adapted to the demands of the environment. How could evolution shape something as amazing as the human eye, surely it must have been designed? I couldn't get my head around the deep time that was required to produce the vast array of flora and fauna from what must have been, way back, a single source. Mostly though I had an indoctrinated sense that mankind was on a road that was opposite to evolving – what is the word - not devolving? What I am reaching for is a sense that mankind is in a state of entropy, slow disintegration, which is the complete opposite of the idea of evolution.

I was brought up with the belief that man was created in the image of God. Therefore we were all pale reflections of the real thing and seemingly getting further away from the Maker.

So I was held back initially from grasping Darwin's brilliant idea by a powerful combination of religious indoctrination and a lack of imagination. However, I am no longer a skeptic. Evolution happens and random gene mutations are critical for non-random selection.

Yet, I have a confession. This is not easy to summarise, as I don't want you to think that I am slipping back into a false assumption that all humans, and all other living things for that matter, are a poor copy of some essential and immutable original in the sky. But something very important doesn't add up in the theory of evolution and I will address this in Chapter 14.

I felt compelled to study the theory of evolution and genetics because the ideas made important contributions in the free will versus determinism debate but I was attracted to what I considered to be a much more interesting set of ideas that starts by seeing mankind as a categorising species.

George Kelly (1905-1967) an American psychologist, mathematician and aspiring engineer produced a system that used the category as the basic unit. His theory of personal constructs, first published in 1955, explains why is it that two people seemingly in exactly the same situation behave in different ways. The reason is, of course, that they are *not* in the same situation. Each of us sees our situation through the goggles of our personal construct system. Central to Kelly's theory is an understanding that persons differ from each other in their construction of events.

Kelly argues that it is useful to see constructs as having two poles; a pole of affirmation and a negative pole. Most people recognise bipolarity when it has an explicit verbal label to cover it; black versus white, up versus down, nice versus nasty and so forth. However, Kelly asserts that even when there is no label readily available for the contrast, we do not affirm without implicitly negating within a context. There would be little point in asserting that 'I'm tired' if the contrast assertion of freshness and energy were not implicitly around somewhere to be negated.

Life can be defined as the activity of movement between polarities. People find themselves moved and motivated by the forces created in the tension between these extremes. The poles of life are indeed like the positive and negative poles between which an electric current passes. Existence is the tension between life and death and many other constructed polarities. Without this tension that involves us in continuous aspirations and desperations and constant ups and downs there would be no human existence at all. Life takes place in the force field created by the process of affirmation and negation.

I liked the theory of personal constructs because it sees the individual as an inveterate inquirer. In this we are self-invented and shaped, sometimes wonderfully and sometimes disastrously, by the direction of our enquiries.

The school of psychology that Kelly's theory gave rise to recognised that psychology is our understanding of our own understanding. This contrasted sharply with much of what I learnt on my undergraduate programme, which incorporated a great deal about rats and pigeons. By placing the individual as the inquirer it presented a framework embodying the kind of systematic thinking and experimental articulation, which are the universal aspects of science whilst at the same incorporating the wisdom of philosophers.

I particularly liked the key idea, made explicit in Personal Construct Theory, that we cannot contact an interpretation-free reality directly. We can only make assumptions about what reality is and then proceed to find out how useful these assumptions are. Whilst this is a relatively popular contention put forward in philosophy and literature, psychologists tend to revert to the notions of a reality whose nature can be clearly identified. Hence the use of the term 'variable' as in the phrase, 'variables such as intelligence must be taken into account'. I can hear Kelly shouting to us; 'Intelligence is a dimension which man has invented and in terms of which he can construe others'. Intelligence is not a thing that must be taken into account. Once invented it is difficult to 'uninvent' something, but it is possible to use entirely different constructs that do not involve such a dimension at all.

Kelly identifies the fallacy in behaviourist psychology (and its sophisticated derivatives) that we simply respond to a stimulus. We respond to what we *interpret* the stimulus to be and this in turn is a function of the kind of replications (constructs) we have detected in, or imposed upon, our universe. This principle can be extended to all living organisms. Experimental animals do not simply respond to whatever the experimenter may choose. It is more like the other way around. The experimenter must, first of all, find and select items that the rat or pigeon (or human for that matter) will look for, rather than merely perceive, and will choose to respond to. This fact underlies the well-known and insightful cartoon in which a pigeon in a laboratory is telling a newcomer that all that they have to do is press a lever to get a little old man to send in some food. That story is made more real for

12

me as when I typed the sentence my cat meowed and I immediately jumped to attend to its needs!

In contrasting his approach with that of Freudians (you are the victim of your infancy) and behaviourists (you are the victim of your reinforcement schedule), Kelly argues that an individual is never a victim of their autobiography although they may enslave themselves by adhering to an unalterable view of what their past means.

Construct theory was deliberately stated in very abstract terms to avoid, as far as possible, the limitations of time and culture. The central tenets of the theory are stated in the form of a fundamental postulate and eleven corollaries. These are clearly detailed in Kelly's 'A Theory of Personality – The Psychology of Personal Constructs' so I will not repeat them here. However I will cover a very important implication.

Kelly implies that we are not reacting to the past so much as reaching out for the future. As such, we check how much sense we have made of the world by seeing how well we anticipate it. The word 'anticipate' is nicely chosen because it links with the idea of prediction; of reaching out and being an individual agent in the process of constructing the future.

Can I bring you back to my teenage interest in the free will versus determinism debate? Kelly's theory had clear implications. One of them is that 'free-determined' is a construction that we place on acts and is only useful to the extent that it discriminates between acts. To say that humans are entirely determined is as meaningless as to say that they are entirely free. The construction (like all our interpretations) is useful only as a distinction and even then the distinction must have a specific range of convenience. A person is free with respect to something just as he or she is free in respect to something else. In this way construct theory avoids the deterministic argument that puts the arguer in the paradoxical position of being a puppet deciding that he is a puppet.

With the passage of time, it is apparent that the heredity versus the environment debate that my university professors were interested in was doomed to rattle on unless it was recast using different distinctions. The utility of the polarity, heredity-environment had run its course. It is clear now that the controversy over whether these factors or some interaction between the two causes behaviour is just incoherent and I will expand on this in Chapter 13.

What was the impact on me?
During my undergraduate years I was disappointed with the power of prayer. Apparently the boundaries of divine intervention are drawn very tightly. On the other hand, it became clear to me that freedom is built into the process of living and is absolute in the sense that

13

whatever you choose to do is always fresh and new. It is so because all organisms live in an open system, one in which nothing actually repeats itself entirely.

'There are those who imagine the unlucky accidents of life – life's 'experiences' - are in someway useful to us. I wish I could find out how. I never know one of them to happen twice. They always change off and swap around and catch you on your inexperienced side.' Mark Twain

It is interesting to reflect on this lack of repetition in the living world. Everything is a one-off in the sense that no particular choice will occur again. Each choice decision is specific to the circumstances of that particular moment in time, including our history encapsulated in memory, our state at that particular moment which will never be the same again and the construction of the external surround that never will be quite the same again. Taking all this into account there is a unique set of circumstances within which living organisms make their choices. No wonder free will had to be invented! Without it we wouldn't be able to cope with the gap in knowledge that life throws at us.

By the end of university I was clear about something that all my friends had been clear about from the very beginning and had not wasted time and effort to dig too deeply into, namely that we are all free agents.

I finished university with a teenage interest in free will satisfied. What has struck me since is that many people seem to use their freedom only for the purpose of denying its existence. Highly talented people seem to find their purest delight in magnifying every mechanism, every inevitability, everything where human freedom does not enter or does not appear to enter. A great shout of eureka goes up whenever anybody has found further evidence in physiology, psychology, sociology, economics or politics that people cannot help being what they are or doing what they do. The denial of freedom is of course a denial of responsibility. There are no acts only events; everything simply happens, no one is responsible.

Chapter 2
Transition

At University College London I was interested in the question, how much space have I got? When I started at London School of Economics (LSE) it was clear that Governments were interested in how much space they have to make a difference.

Now I want to flag that we are taking a slight detour on our journey. This is for the purpose of demonstrating that underlying any system is a belief about the nature of uncertainty.

I started at the LSE in the academic year following the Winter of Discontent which is an expression, popularised by the British media, referring to the winter of 1978–79 in the United Kingdom, during which there were widespread strikes called by local authority trade unions demanding larger pay rises for their members. The weather turned very cold in the early months of 1979 with blizzards and deep snow. It became the coldest since 1962–63, which added to people's misery.

Mirroring the discontent manifested on the streets, the world of politics and economics was fomenting. At the heart of this was a mighty tussle that can be simplified as the big fight; Hayek versus Keynes.

Let's consider in broad-brush terms the thinking of Frederick Hayek (1899-1992) and John Maynard Keynes (1883-1946). In a nutshell, the big fight billing was this; Keynes was telling politicians that intervention by policymakers could make things better whereas Hayek was saying they would only make things worse.

Hayek as a young child lived through the hyperinflation in Austria in the 1920s. This must have influenced his view that the only government power he had confidence in was the power to make things worse by debauching the currency.

Hayek became the great free-market thinker who argued with Keynes in the 1930s over government intervention in the economy. One analysis of the Great Crash of 1929 is that financial markets had become too free. Hayek's analysis was exactly the opposite. In his view, it happened because the markets weren't free enough. In summary, the argument that Hayek made was that the 'Fed' caused the crash by meddling and the worst kind of meddling, Hayek thought, came in the government's determination to control the price of money, which it did through the interest rate. By keeping interest rates too low it encouraged a lot of poor investment in projects or assets, which were not economically worthwhile. Hayek's analysis was that meddling might stave off a more serious downturn but only at the cost of encouraging people to take on debts they couldn't afford and giving banks an incentive to take excessive risks. He also said that further efforts to stimulate the economy would only make things worse, especially if that meant more borrowing by government.

Shortly after World War II Hayek wrote a best-selling polemic, *The Road to Serfdom* in which he railed against economic planning. In it, he warned that the dead hand of the bureaucrat could threaten a free society almost as much as the iron boot of Stalin. After that, Hayek had years in the intellectual wilderness until the 1970s when there was a last burst of fame. He was awarded a Nobel Prize for economics and feted by free-market politicians on both sides of the Atlantic.

Hayek's call for policy makers to leave the economy well alone was due to his appreciation of the great complexity of markets and their inherent unpredictability. Hayek was clear that policymakers couldn't master those complexities well enough to guide the economy in the right direction.

Let's consider the other person on the big fight billing. It's hard to think of a British man born in the 1880s whose name you hear more often, in current economic debates, than John Maynard Keynes. Keynes has gone down in history as the first of the big spenders; the man who encouraged governments to spend their way out of problems.

Like Hayek, Keynes was influenced by what he saw in his formative years. He understood that countries were inter-dependent and if you beggar your neighbour (and trading partner) you might very well beggar yourself. Keynes saw the world learn that lesson the hard way in the years after World War I. When he personally helped the Allies create institutions like the World Bank at the end of World War II he was trying to make sure we wouldn't get into such a mess again.

Keynes was one of the first to say in the 1930s that economies could just get stuck; they might sink and not float magically back up. When that happened his solution was for the government to borrow its way out of trouble. However, he also said governments should balance the books in the good times.

Keynes' basic idea that government intervention was necessary when things go wrong was gradually built into the fabric of every Western economy from the 1940s onwards. Yet, Keynes, just as Hayek, felt that the world was a deeply unpredictable place and uncertainty an inescapable part of economic life. Keynes said, in effect, that you should never think you have abolished boom and bust. This view was reinforced when, at one point, his own investments nearly bankrupted him. As such, whilst Keynes might be the grandfather of activist government, he certainly gave policy makers cause to reflect.

The man who did most to make economists and politicians believe that

16

they could bend the world to their will also gave them the very best reasons for self-doubt. Stephanie Flanders

What was the outcome of the clash of ideas in the late 70s? Margaret Thatcher and Ronald Reagan, both fierce advocates of free markets, backed Hayek in the blue corner. It is reported that Margaret Thatcher would pull favourite Hayek quotations from her handbag at key moments during cabinet meetings. However, Hayek never landed the knockout punch. The full-blown ideas were too much to handle. It was like Turkeys voting for Christmas; politicians could not give up that much power to intervene and meddle so they went with the half way house – monetarism, and the guru that they turned to on the subject was Milton Friedman.

What this slight detour shows is that the greatest of economists were of one accord. The world is a deeply uncertain place.

It is an idea from a much less well-known economist in the 1970s that, for me, summarises just why Hayek and to a lesser extent Keynes had uncertainty as a main element in their thinking. George Lennox Sharman Shackle (1903-1992) argued that imagination is what people must 'substitute for knowledge in that vital and limitless area where we are eternally denied it, 'tomorrow''. If tomorrow is a figment of our collective imagination, it is not surprising that it can turn out to be very surprising and deeply unpredictable.

The Big Ideas

Although the Hayek versus Keynes battle was a backdrop to my studies in personnel management and industrial relations the big ideas that grabbed my attention came from an economist, E F Schumacher and a social scientist, Elliott Jaques.

At the LSE at the time I had as a personal tutor, Chris Schumacher, the son of E F Schumacher of *Small is Beautiful* fame. As such it was incumbent upon me to read and digest the book. I have reread it recently and marvelled at the insight, wisdom and prescience of the author. I have summarised below the points that stood out for me I as ventured into the world of industry and commerce; one that was rapidly about to change in the United Kingdom.

Schumacher wrote that all real human problems arise from the antinomy of order and freedom. Antinomy means a contradiction between two laws; a conflict of authority; opposition between laws or principles that appear to be founded equally in reason.

Here we have another of these polarities encountered in the previous chapter. Again, life happens as a result of the interplay of these tensions.

Schumacher expands on this antinomy. In any organization, large or small, there must be certain levels of clarity and orderliness. If things fall into disorder nothing can be accomplished and all is noise, bustle and waste. Yet orderliness, as such, is static and lifeless. For progress and change there must be plenty of elbow-room and scope for breaking through the established order to do the thing never done before, never anticipated by the guardians of orderliness.

We can associate many pairs of opposites with this basic pair of order and freedom. Centralisation for instance is mainly an idea of order; decentralisation, one of freedom. Order requires intelligence and is conducive to efficiency; while freedom calls for, and opens the door to, intuition and innovation resulting in a new, unpredicted and unpredictable outcome.

The larger the organization, the more obvious and inescapable is the need for order. However, Schumacher is clear that if this is attended to with such efficiency and perfection that no scope remains for disorder, the organization becomes moribund and a desert of frustration. It becomes more and more efficient until the day that it is bypassed by creative ideas from elsewhere. Thereafter, efficiency counts for nothing, as the enterprise is no longer effective.

Excellent, I thought! Schumacher talks of real life, full of antimonies and bigger than logic. Without order, planning, accountancy, instructions, targets, obedience, discipline, systems and processes nothing fruitful can happen because everything disintegrates. And yet, without the fresh air of disorder, the creative spark, the entrepreneurship venturing into the unknown and unknowable life is entirely repetitive and predictable.

Schumacher was sure that the centre can easily look after order but it is not so easy to look after freedom and creativity. The centre has the power to establish order but no amount of power evokes the creative contribution.

How, then can top management at the centre work for progress and innovation? Assuming that it knows what ought to be done, how can management get it done throughout the organization? How can managers create the space for the individual to be creative? These were the questions posed for me by Schumacher (with his ideas on 'intermediate technology' plenty of others received different questions and answers). They were questions left hanging in the air for many years as it seemed that management science, organizations and institutions largely focus on the order side of the order-freedom polarity.

The other person that struck me as having a great deal to say was Elliott Jaques (1917-2003). His work was beginning to become known and the ideas struck me then as deeply sensible.

Jaques, who defines work as 'an organism's use of judgement in making the decisions necessary to reach a goal', identified that attitudes and feelings about work are very different from the

18

dictionary definitions. He contended that everyone needs to be engaged in socially valued work and that employees resent senior management's failure to understand how deeply seated the importance of work is to them. They want work that provides them with the opportunity to exercise their full potential, to receive fair compensation and to not be subjected to the usual 'artificial carrot-and-stick treatment' used to coerce them into doing their jobs.

This was my first acquaintance with the ideas of Jaques and I have subsequently spent much time thinking about his work and applying general principles from it. I have written extensively about Elliott Jaques in *It's About Time* and will not repeat his life story here. However, the title of my book provides the clue to one of Jaques major influences on social science and management theory. Jaques identified time, specifically the time factor that relates to intention, as the means to objectively measure both the size of jobs and the capability of people to cope with specific levels of challenge.

There is more on Jaques to come as we meet him later when his ideas have fully matured.

What was the impact on me?
Economic ideas had the biggest direct influence on me. I left the LSE just as the Conservative Government's policy began to bite into what it could only have thought was the soft rump of British manufacturing industry. Although jobs were incredibly difficult to find in the recession of the early 80s I managed to join British Aluminium as a Graduate Trainee. British Aluminium was part of Tube Investments, which at the time made a range of products from Raleigh bikes to Russell Hobbs kettles. By the end of the decade the 'metal-bashing' base within Tube Investments had largely disappeared and the company itself was merged into Smiths Industry in 2000. Tube Investments was not an isolated case. The country that gave rise to James Watt and Isambard Kingdom Brunel, it seemed, just lost interest in making stuff. It is still a marvel to me how quickly de-industrialisation happened in the UK. In the wider economy, jobs were migrating to the service industries including retail and high technology.

It is interesting for me to reflect on the rise of high technology and I am reminded of the book by Alvin Toffler, *Future Shock.*

'The illiterate of the 21st Century will not be those who cannot read and write, but those that cannot learn, unlearn and relearn.' Alvin Toffler

In 1970, the book gripped America and some of Toffler's predictions are still unfolding. For Toffler, life was changing at a faster and faster rate, everything from technology to family structure through to

19

politics. The result was a kind of culture shock of the future, 'too much change in too short a time'. In *Future Shock*, 'the future arrives too soon and in the wrong order'.

Inspired by his book, a group of fellow students sat around trying to think of what the future will throw at us. As the outputs of the thinking were part of our coursework the results were typed out (on a manual typewriter), acetated and presented on an overhead projector.

Despite the group comprising some pretty smart cookies with a grasp of technology not a single person voiced a thought that in a few years the means of delivering our task would have changed dramatically. PCs and photocopiers in every office were the stuff of science fiction. Even science fiction hadn't imagined the next generation of computer-based systems, tools and applications. The world-wide-web was still only in Spiderman comics! That's how fast the world was coming at us.

As I ventured out into the world of work that had seen the rise of the mainframe computer but not the distributed systems that were to come, I retained, at the back of my mind, the thoughts generated from reading *Small is Beautiful*. Specifically I kept in mind that everything that management did seem to be an attempt to bind and control people. Yet, it was like trying to put one's thumb in the hole of a breached dam. The creativity came through anyway but because it was often resisted, the ideas tended to sweep away what was there. I kept on looking for a system that promoted a balance of order and disorder and I address this later in the book.

Chapter 3
The Digital Revolution

With the decline of manufacturing industry and the growth of information technology, it wasn't long before I jumped on the rapidly accelerating high-tech gravy train (we all know what happens to gravy trains!).

I have spent the majority of my career in the high-tech/computer/telecoms world and frankly it was a great time before it hit the buffers! That's what happens to gravy trains and I'm sorry about the messy metaphor.

Most of my time, I was learning the trade of HR but I won't focus on this - you may be pleased to note.

I had the opportunity to work in the computer industry with some real boffins and they provided a background and early exposure to the groundbreaking idea that mathematics has got a lot to say about human nature.

Right up to the middle of the twentieth century life was thought to be qualitatively beyond physics and chemistry, (subjects to which mathematicians have much to contribute). No longer. The difference between life and non-life, my boffin lunch-mates informed me, is a matter not of substance but of information. Living things contain prodigious quantities of information. Most of this information is digitally coded in DNA and there is also a substantial quantity coded in other ways.

Before elaborating this though, a caveat! I am, in old terminology, a Friday afternoon production model. Just as old British cars made before the weekend were supposed to be riddled with defects, a major component in my brain was missed out along the production line. Whilst very happy in the world of words, I have a blind spot for numbers. So forgive me if I open up a fascinating area but don't venture in too far.

The Big Ideas

George Boole (1815-1864) in his work entitled the *Laws of Thought* clearly linked mathematics with the fundamental operation of the human mind.

The design of the following treatise is to investigate the fundamental laws of those operations of the mind by which reasoning is performed; to give expression to them in the symbolical language of a Calculus, and upon this foundation to establish the science of Logic and construct its method; to make that method itself the basis of a general method for the application of the mathematical doctrine of Probabilities; and, finally, to collect from the various elements of truth brought to view in the course of these inquiries some probable intimations concerning the nature and constitution of the human mind. George Boole

George Boole created a schema of four logical types (which replaced Aristotle's two logic types). These deal with the logical truth of 'or-or', 'and-and', 'if-then' and 'if-and-only-if' propositions depending on the hypothesized truth or falsity of their elementary statements. In slightly more detail, these four logical operators (sentential connectives) are: 1) the disjunctive 'or' (where the choice is between *this or that*), 2) the conjunctive 'and' (where the choice is made by putting together *this and that and the other*), 3) the conditional 'if-then' (where the choice is made through a logical sequence of *if A then B and if B then C*) and 4) the bi-conditional (in which B is chosen *if and only if there is also a C*). Boole understood that he was dealing with some very basic psychological processes. His laws of thought were the foundation for another genius mathematician, Alan Turing (1912-1954) to address the question of what goes on in the head of the human computer. Turing realised that although the human 'mind' is complex and conscious, capable of imagination and insight, these capacities are not required to compute something. Only a set of rules had to be followed precisely.

Turing showed through his logic that you could have a mechanical (unthinking) process to perform the same act as achieved by a higher level of intelligence. Effectively he removed the necessity of the human agency with all its higher processes to perform computations.

Turing saw that calculation had two aspects, data and instruction on what to do with the data. From this he imagined a machine that could do anything, through feeding it enough information and instruction. This is the Universal Turing Machine.

As you probably know, Turing's ideas have become the basis of modern computer science but they also form the bedrock of cognitive psychology.

Whilst Turing's ideas were essential to the meteoric rise of computing it needed others to shed light on to the fundamental nature of information before the computer could become of age. Key to this is another mathematician, Claude Shannon (1916-2001). Shannon was working on a problem for Bell Telephone Company. Each day Bell relayed vast amounts of information but in its analogue form, it could not be measured. In effect it had a flawed model. It did not understand the main ingredient of its business.

Shannon took the vague notion of information and pinned it down. He realised that information was related to how unusual it was. 'Man bites dog' is news because it is unexpected. He reduced information to

22

a fundamental element; to a binary digit. The primary unit of information is contained in anything that is capable of being only in two states, on or off, stop or go, heads or tails, 1 or 0. This primary unit Shannon called the 'bit'.

Identifying the bit helped launch the digital revolution and information theory. Information theory identifies that information is not something that only humans create and that it doesn't just describe content. Rather, information is an inseparable part of the world. Information is embodied.

I take this one step further. What is '1 or 0', 'stop or go', 'heads or tails' if it is not a choice? Choice, as embodied in the bit, is the fundamental atom of life.

Whilst you ponder on the last statement let me be clear that 'choice' should not convey a sense of something that must be conscious and capable of verbal articulation. It is a major point in this book that all living organisms make choices whether or not they can verbalise these. More of this later!

What was the impact on me?
Boole, Turing and Shannon's ideas have had a huge impact on my life and that of everyone else for that matter. It shows how fundamental ideas start off small and apparently insignificant, coalesce and have their main impact well into subsequent generations and in many cases well after the originators have passed away. In the case of Turing it is common knowledge that he did not see the main impact of his work.

Turing's homosexuality resulted in a criminal prosecution in 1952 when homosexual acts were still illegal in the United Kingdom. He accepted treatment with female hormones (chemical castration) as an alternative to prison. Turing died of cyanide poisoning in 1954, just over two weeks before his 42nd birthday. If we think that everything is getting worse in society, we only need to recall the personal tragedy of Alan Turing to realise that we are very much mistaken.

Before moving on in our tour, it is interesting that the Conservative Government of the UK stepped up its level of privatisation of public companies on the back of the perceived success in shaking up the telecommunication industry. Yet, it is clear that telecoms was shaken up because of the influence of the ideas of Boole, Turing and Shannon to name just three that enabled the computing and digitalization revolution. It was not shaken up because shareholders were given the right to unearned income! The same coalescence of ideas had not happened in the water, power-generation or railway industries and, as it turned out, privatisation has added to costs and delivered only marginal increases in services. To my way of thinking, these industries could have remained nationalised (especially as they still demand huge government subsidy) without any significant drawback.

This venture into the realms of politics is to provide an example of a human bias that we will meet again in due course. This is a bias to see cause and effect everywhere (e.g. privatisation caused the invigoration of the telecommunication industry) and we make this connection even if there is none. Levitt and Dubner in their best seller, *Freakonomics,* provide plenty of examples where cause and effect have been wrongly attributed.

Privatisation of utilities had another impact. In the mid-eighties we were all encouraged to be investors and I, like everyone else, thought that being a Sid (the marketing terminology at the time for a small investor) would be good for me. Fine in theory but in practice whenever I have needed to cash in the purchased shares they have invariably been well below what I have paid for them; a track record that is on a par with some of our most reputable investment institutions.

I have not mentioned my abysmal investment record just to get this off my chest. The stock market is a deeply uncertain place and has proved to be an excellent real life laboratory to see a range of inbuilt human biases at work. We will return to these later.

Chapter 4
Imagine '97

In 1993 Mercury Communications began the biggest experiment in organizational management ever conceived.

Mercury Communications was a national telephone company in the United Kingdom. Formed in 1981 as a subsidiary of Cable & Wireless, it was set up to challenge the monopoly of British Telecom (BT) and by 1990, which was when I joined, was providing a full range of voice and data services.

Mike Harris was the CEO for a time in the early part of the 90s and he was clear, just as other senior figures in the industry were, that convergence of computers and telecoms provided huge opportunities. Rather than starting as many organizations were doing with a technical strategy and driving this top down, Harris created a different kind of vision; one that had an acronym, 'P.I.E. in the palm'. This together with the name of the change programme, 'Imagine '97' was emblazoned on tee shirts provided to all employees when they attended large group sessions.

Within three years (by 1997) Harris proclaimed, 'Mercury will be a world leader, providing **p**eople, **i**nformation and **e**ntertainment (the pie bit) to a device in the palm of the hand'. This was years before the iphone. It was entirely possible although Harris and his senior team at that time had no idea how it would be achieved. They put their trust in the power of the word and that by vocalizing their ambition the possibilities would open up to realise this. Not only was trust inherent in the strategic process, Harris made it part of the fabric of the organization. He empowered people to 'declare a breakdown' when they found company politics getting in the way of delivering better, customer focused processes and innovation.

The great experiment never ran its course. There were many entrenched views in the organization about what made for success and that didn't include giving the right to the call-centre employee, for instance, to stop the process. Also the vision was hype. Of course it was; it certainly wasn't an extrapolation from current trends and the paymasters in the parent company didn't like unfounded ambition. It was deemed that Mercury had lost its way and had become inefficient. Harris was moved and it took all of a few months for everyone to calm down and carry on.

The ghost of Mercury lives on bizarrely in the Mercury Music Prize. Imagine what it would have been like for Britain PLC if Mercury had beaten Apple to the punch.

Big Ideas

Imagine '97 was a big idea and it introduced me to other big ideas. One of the quotes used in the programme was the philosopher, Heidegger's, 'language is the house of being'.

Heidegger's quote certainly was enigmatic and intriguing. This, together with the whole of the Imagine '97 programme, made me think and read about the power of language. In particular I have been influenced by the ideas of Noam Chomsky whom I first studied at University College London. Honestly, I struggled with him then but through the writing of Professor Stephen Pinker, his ideas have made a great deal more sense.

People are passionate about language. Language makes humans distinct in the animal kingdom and we want to know about language because this knowledge, it seems, will lead to insight about human nature.

The human eye is often cited as the most amazing of the human organs. However the language 'organ' is perhaps even more intricately constructed and is composed of many parts including syntax with its combinational system building phrase structures, phonological rules and structures, speech perception, parsing algorithms and learning algorithms. These are physically realised as the ability to dispatch an infinite number of precisely structured thoughts from head to head by modulating exhaled breath. Wow!

Human language is not just about communication. It is about the creation of permanent knowledge by means of articulating that knowledge in language and placing that language firmly and permanently out into the tangible world so that it suffuses that world for all to hear (and from about 5000- 3000 BC to read) and to use.

How we see things is not only a matter of what the tangible circumstances are but also how we 'linguify' them. The rather obvious fact opens up the possibilities for all kinds of constructive associations and, at the same time, for all kinds of misleading constructions of what might be going on. The dangerous part of the idealisation of language is that it plays tricks with our sense of reality. Just because we have a word for something it is assumed that that something really exists. The 'languaged' world comes to be seen as the real world.

Language sets us apart from the rest of the animal kingdom; it is our highest capability. As such, it is difficult to think that it might be an instinct.

Darwin himself first articulated the conception of language as a kind of instinct in 1871. In the *Descent of Man* he had to contend with language because its confinement to humans seemed to present a challenge to his theory. He concluded that 'language ability is an instinctive tendency to acquire an art'.

26

In the last century, the most famous argument that language is like an instinct comes from Noam Chomsky, the linguist who first unmasked the intricacy of the system and perhaps the person most responsible for the modern revolution in language and cognitive science.

Chomsky provided two fundamental facts about language. Firstly, virtually every sentence that a person utters or understands is a brand new combination of words. As such, a language cannot be a repertoire of learned responses. He argues that the brain must contain a recipe or program for building an unlimited set of sentences out of a finite list of words. This program can be called a mental grammar (not to be confused with stylistic grammars which is the etiquette of written prose). Secondly, children develop these complex grammars rapidly and without formal instruction and grow up and give consistent interpretations to novel sentence construction that they have never encountered before. Therefore, Chomsky argued, children must be innately equipped with a plan common to the languages of all languages, a 'Universal Grammar', which tells them how to distill the syntactic patterns of the speech of their parents.

By performing painstaking technical analyses of the sentences ordinary people use as part of their mother tongue, Chomsky and other linguists identified the mental grammars underlying people's knowledge of specific languages and of the Universal Grammar underlying these specific grammars.

Reactions to Chomsky ranged from awe-struck deference to instant dismissal. This, in part, is because Chomsky's thoughts broke the paradigm of the age. He attacked what was one of the foundations of the twentieth-century's intellectual life; the Standard Social Science Model.

Relativism began to dominate intellectual life in the 1920s and gave rise to what has become known as the Standard Social Science Model. This was based on a fusion of ideas from anthropology and psychology and included the belief that human behaviour is determined by culture and that learning is a general-purpose process involving reward/punishment and the following of role models.

The Standard Social Science Model has not only been the foundation for the study of humankind within most American institutions for a large period of the last century but it still serves as the secular ideology of our age; the position on human nature that any decent person should hold. Coincidentally it all fits very nicely with the American Dream. The alternative, sometimes called 'biological determinism' is said to assign people to fixed slots in the socio-political-economic hierarchy and to be the cause of many horrors of recent centuries such as slavery, colonialism, racial and ethnic discrimination, economic and social castes, forced sterilization,

27

sexism and genocide. Who in their right mind could stray from the path of the Standard Social Science Model?

The theory that language is based on using abstract variables and data structures was a shocking and revolutionary claim to those working within the Standard Social Science Model as the structures have no direct counterpart in the child's experience. Some of the organization of grammar would have to be there from the start; part of the language-learning mechanism that allows children to make sense out of the noises that they hear. Language is a beguiling case illustrating that our ability to handle complexity is not caused by learning; learning is caused by our ability to handle and process complexity!

Professor Steven Pinker at MIT argues that complex language is universal because children actually reinvent it generation after generation. This isn't because they are taught, nor because they are generally smart nor because it is useful to them but because they just can't help it.

When it comes to thinking about thinking it is easy to think that thought is the same thing as language. However, with a moment's reflection, we recognise that there are occasions when we cannot find any words that properly convey a thought. Yet, we can be forgiven for equating words with thought. As Professor Pinker points out in *The Language Instinct*, thoughts are trapped inside the head of the thinker and to know what someone else is thinking or to talk to each other about the nature of thinking we have to use, yes, you have guessed it, words!

Although in the past thinking has largely been taken as the same thing as language, cognitive science is breaking the word barrier and assessing many kinds of non-verbal thought. In addition, there is a theory of how thinking might work that formulates the questions in a mathematically precise way. Together these developments are enabling us to think about thinking in a new way.

The experimental studies that assess many kinds of non-verbal thought involve babies (who cannot think in words because they have not learned any), monkeys (because they are incapable of learning them) and some human adults who claim that their best thinking is done without them. This evidence builds into a clear picture that images, numbers, kinship relationships and logic can be *represented* in the brain without being couched in words and can be manipulated through a non-verbal 'mentalese'.

In the first half of the 20th century philosophers struggled with the idea that thoughts are representations and concluded that reifying thoughts as things in the head was a logical error. A picture or family tree or number in the head would require a little man (a homunculus) to look at this. And what would be inside his head? Well, even smaller pictures and an even smaller man looking at these pictures.

It took Alan Turing (yes we are back with him) to make the idea of mental representations scientifically respectable. Turing described a hypothetical machine, a Turing Machine, which could be said to engage reasoning. In fact, this machine is powerful enough to solve any problem. It uses an internal symbolic representation, a kind of mentalese. By looking at how a Turing Machine works it is possible to get a grasp of what it would mean for a human mind to think in mentalese as opposed to English or any other language for that matter.

Using Turing's ideas, cognitive scientists developed a theory of thinking that is called the physical symbol system hypothesis or the computational or representational theory of mind. It is as fundamental to cognitive science as the cell doctrine is to biology and plate tectonics is to geology.

It could be argued that mathematics and cognitive science reduces man to a machine but that would belittle its contribution. Turing showed that complex calculations could be done without the sophisticated components, such as language and consciousness, that people bring to solve problems. Therefore, it is entirely possible that amoebae for instance are capable of some primitive representations. Turing does not deny the aspects that we call human. What he does show, and this seems to be the same as all great insights, is that we are directly connected to the rest of the animal kingdom and to all living organisms.

By taking language as an instinct it enables the study of language to become part of evolutionary psychology and the search for a universal design. However, this goes completely against the Standard Social Science Model.

Underpinned by the doctrine of Relativism, the guardians of the Standard Social Science Model felt that by valuing diversity they were ensuring equality. Paradoxically, by focusing on differences, there is less of a sense of a family. Evolutionary psychology focuses on what is shared and, when we look at what it is to be human, we see that differences are of relatively minor interest. Cognitive science and evolutionary psychology build a sense of unity.

This sense of unity is underpinned by genetic evidence. Researchers (Bodmer and Cavalli-Sforza) have identified that eighty five percent of human genetic variation consists of the difference between one person and another within the same ethnic group, tribe or nation. Another eight percent of the difference is between ethnic groups and a mere seven percent is between races. In other words, race, it appears, is quite literally skin-deep. However, to the extent that perceivers generalize from external to internal difference, nature has duped them to think that race is important. The X-ray vision of the molecular geneticist reveals the unity of the species.

What was the impact on me?
Although Imagine '97 hardly got off the ground, it had its impact on me. I was filled with boldness; a sense that life would take me to exciting places so long as I was prepared for turbulent water and ready to paddle hard.

The big ideas were put to the back of the mind waiting, it turns out, to coalesce with a range of soon-to-be discovered ideas.

Chapter 5
It's the System Stupid

In February 1994 I started working in Riga, Latvia for the recently privatized telecoms company, Lattelekom that my employer had invested in.

It was less than four years after the fall of the Soviet Union and life was hard for the inhabitants of this former satellite state. If communism had been tough to live with, its collapse was devastating for a time. Factories shut virtually everywhere and there was no social welfare system to really cope. Latvia and the other Baltic States were in suspended animation. Communism had ended but capitalism hadn't really got under way unless you counted the huge black market and mafia extortion as part of this.

When I arrived in temperatures of minus 15 degrees it was like being on the set of a remake of 'From Russia With Love'. People shivered in fur hats in front of newspapers that were posted outside the government buildings, Moskvitchs were the main cars, champagne and caviar were cheap and buying food in the shops involved a curious process of ordering, buying a ticket and collection.

All of that was about to change as a huge tide of dirty-money, EU funding and inward investment was about sweep through the Baltic States. However for a time, there was a collective holding of the breath. For me, it was all new, yet somehow strangely familiar. It was a world largely without personal computers and mobile phones. There was no overt consumerism and mass tourism had not yet created an inauthentic pastiche of global hotel brands, fashion chains, coffee emporia and souvenir vendors. Yes, for me, it all had the vague sense of being back in the 1960s.

Open-plan working had not yet arrived and the head office was full of small offices. When entering these there was always a radio on. I liked the cosy atmosphere that this created but it appeared that there was a sinister reason for the radio's drone. The noise, I was quietly informed, made any bugging ineffective. I was not sure whether we, the newly arrived foreigners, were having our legs pulled when told this but it certainly made me talk with hushed tones just in case old habits died hard.

One of the first initiatives that we held in the Company was a 'let's get to know each other' event. Using a facilitator, many of the senior team came together and talked about who they were and what their hopes and fears were for the joint venture. As English was not widely spoken the facilitator encouraged us to draw pictures on flip charts to create a collective vision of the future. I remember how someone drew a beehive to represent, I guessed, hard work and endeavour. Then someone else put the word 'zubb' by it. The Brits were all confused and we asked the translator for assistance. At first she looked as if we were rather simple, 'zubb, you know, zubb, the sound of bees'. I knew at that moment that if the bees in this country didn't go buzz we were really in for a culture shock. And to some extent, so it proved.

If you have ever lived abroad you will know that one of the great things about it is that there's not only a new culture to understand, there is also the opportunity to look back on your own and see its great strengths and weaknesses.

Latvia was recovering from fifty years of oppression and virtually every ethnic Latvian felt that they had been robbed and imprisoned, unable to take advantage of the great freedom given to those in the West. I was not so sure. Looking across to the West from this vantage point I saw that a lot of this, so called, freedom was about making choices to saddle yourself with debt that took a lifetime to repay with shopping therapy exacerbating the problem.

However, I was never allowed to become more than mildly disaffected with the system in which I had been raised. Neither I, nor my family and friends, or anyone for that matter in living memory, had been forcibly deported to live (but most probably die) in freezing conditions without rations or adequate clothing.

Big Ideas
Whilst academics thought about how human behaviour is impacted by such factors as the environment or heredity, there were thinkers that saw human behaviour as being a consequence of the system in which they operated. One such man was Karl Marx.

Surprisingly, Marx gave a spirited rendition of the power of the capitalist system. He understood both the genius of capitalism and its inherent instability. On the plus side, Marx was clear that the idea of conducting the entire economy on the basis of private greed has shown an extraordinary power to transform the world.

'The bourgeoisie, whenever it has got the upper hand, has put to an end all feudal, patriarchal, idyllic relations and has left no other nexus between man and man than naked self interest... The bourgeoisie, by the rapid improvement of all instruments of production, by the immensely facilitated means of communication, draws all, even the most barbarian, nations into civilization.' (Communist Manifesto)

Marx clearly identified the motor of capitalism, namely the search for profit.

The strength of the idea of capitalism lies in the simplicity of the profit motive. It suggests that the totality of life can be reduced to one aspect, namely profits. The business-person, as a private individual, may be interested in other aspects of life, perhaps even in goodness, truth and beauty but as a business-person they concern themselves only with profit. Capitalism fits neatly with the idea of a market. The market provides a price tag for all goods and services and therefore enables us to measure very different items with a common ruler. A £100 of oil is equal to £100 of shoes, which equals a £100 of insurance. The sole criterion to determine the relative importance of these different goods is the rate of profit that can be obtained by providing them. If any categories yield a higher profit, this is taken as a signal that it is rational to put additional resources into the former and withdraw from the latter.

Everything becomes crystal clear after you have reduced reality to one and only one of its thousand aspects. You know what to do; whatever produces profits. You know what to avoid; whatever reduces profits or makes a loss. At the same time there is a perfect measuring rod for the degree of success or failure.

The real strength of the idea of private enterprise lies in this ruthless simplification. It is no accident that successful businessmen are often astonishingly primitive; they live in a world made primitive by a process of reduction. They fit into a simplified version of the world and are satisfied with it. They hire apprentices that fit their self-image and the system becomes self-sustaining. Business people do not befog the issue by asking whether a particular action is conducive to the wealth and well being of society, whether it leads to moral, aesthetic or cultural enrichment. Business people simply ask whether it pays. The process is simple; investigate whether there is an alternative that pays better. If there is, choose the alternative. However, when the complexity of the real world occasionally makes its existence known and attempts to force upon their attention a different one of its facets they can became quite confused.

Marx looked carefully at capitalism and his analysis was simple, the problem was private property. No private property – no problem. Indeed, Marxism is founded on a belief that, once upon a time, human nature was satisfied because there was a simple relationship between the effort of work and replenishment. However, when people appropriated land and property and made the dispossessed work for them, the workers had to dissociate the pleasure of survival from the work it involves. Marxist analysis views this as alienation. Based on a conception of human nature, Marx went on to preach that getting rid of the 'evil' of private property would return humans to their natural state. Joseph Stalin took that thinking a step further; no people – no problem.

When I went to Latvia the influence of Stalin on the collective psyche was there to see. Whilst there were no statues and overt reminders of the deportees to Siberia, Stalin's dead hand had left its mark and was evident in the huge level of distrust that pervaded society. Yet, paradoxically there existed a genuineness in interpersonal relations that was more real than the emerging Americanized 'have a nice day' culture in Britain.

Living in Latvia at the time of its renaissance was exciting and I tracked the country's progress when I left. It is a lesson in the power of the profit motive. Has it delivered? Absolutely. There are no queues and shopping can be done in high-class malls all over the capital city, Riga. Has it improved living standards? Yes, if measured against the availability of world-class products and services including high-speed Internet access? Has it produced a fairer, inclusive society focused on sustainability and maintaining its unique ecology, cultural heritage and green credentials? No, not really! What it has achieved is the creation of super rich elite and a mass of indebted workers – just like the rest of us.

Would anyone choose to go back to communism based on the Marxist analysis that all our ills are down to private property and that eliminating this returns humans to their natural state? No, of course not! Understanding human nature as essential greedy or nice is a gross misunderstanding of the whole person and gives rise to systems that actually debase to a greater or lesser extent the human spirit.

What was the impact on me?
My stint as HR Director in Latvia had a big impact on me professionally and personally.

In terms of professional development I had to completely rethink HR from the bottom up. Market related pay for instance didn't make sense in a country where labour had had no choice as to where it worked. The market existed but only as a place that you bought potatoes and other staples of life. Performance management did not compute. Standing out, for good or bad reasons, was not encouraged.

Personally, it made me think about my own values and behaviour. I came to realise that trust is made of the emotional equivalent of fine porcelain. One chip and it is highly devalued. Broken, it is almost impossible to repair.

It was a privilege to be present at the rebirth of a nation and from first hand experience I was able to observe how systems radically alter behaviour. However, the biggest impact on me was to leave with a clear sense that a better system must be available for the management of people than the 'best practice' one I had in my professional kitbag at the time.

Chapter 6
Time out

Towards the turn of the Millennium I helped spend my employer's significant cash pile by aiding the purchase and integration of dotcom firms in Europe.

The road ahead was absolutely clear. All of the signposts were pointing the same way. The company's advisors, all very respectable accountancy and consultancy practices, were adamant; voice is dead, long live data; the Internet was the way to go. All the analysts with their spreadsheets shouted the same message; broadband usage was soaring exponentially and the only question remained was how much of the pie (this time, not **p**eople, **i**nformation and **e**ntertainment) do you want? It was the Klondike all over again and there was gold in them there hills.

Asset prices soared as everyone wanted to buy-in and with the Government creating a bidding war for mobile licences it looked like those that were in the telecoms business had endless pockets. Of course, the spreadsheets over-projected the take-up of services and under-projected the costs. The result was the bursting of the dotcom bubble.

The future is never clear; you pay a very high price in the stock market for cheery consensus. Uncertainty actually is the friend of the buyer of long-term values. Warren Buffett.

As it turned out there was gold in the hills but also a lot of useless muck!

Not long after the bursting of the bubble that saw the creation of a large 90% club (membership being easily obtained by having a share price less than 10% of its record price), Gordon Brown, Chancellor at the time mopped his brow. With lowering the interest rate it looked like the effects of the fallout could be contained. He was soon back to claiming that he had eradicated boom and bust but failed to explain that, far from cleaning up the mess, the write-off of assets had been left for all future pensioners to deal with.

For me, the wave of energy that swept through the telecoms industry saw me also borne up and free to do a little of my own thing. I managed to persuade my employer to release me one day a week whilst I studied for a Diploma in Existential Psychotherapy and Counselling. Pretentious, moi! I still haven't thanked enough the HR Director for allowing this.

35

Big Ideas

The diploma was the best course that I have undertaken. Partly this was because I was the only person on it with an HR background and others were grounded in the real world of helping and caring for families and others. The main reason though was that the diploma introduced me to a range of existential thinkers and philosophers.

Existentialism, if it means anything in common culture, refers to a bunch of intellectuals sitting in cafes (often smoking Gauloise cigarettes) having deep conversation about the meaning of life. That is only partly true.

Existential thinking is more than a pastime for arty or intellectual types. It is a steadfast and loyal endeavour to reflect on everyday human reality in order to make sense of it. As a practice it is probably as old as the human ability to reflect. Existential thinkers remind us we can become so engrossed in ourselves that we replace the humble search for meaning with the illusion of obtaining absolute knowledge and mastery. The outputs may lead to technological progress but there is an enormous price to pay. The flipside is the loss of connection with the mysteries of existence.

Existential thinking throughout the history of humankind has arisen in reaction to the dogmatic and pedantic attempts at controlling human destiny. People like Socrates or Jesus of Nazareth can be seen as existential thinkers who set themselves against the brutality and bigotry of their respective cultures of Sophists and Pharisees. At the present time there are still threats to those who would claim their independence of thinking about life. In fact, it is likely that thinking alternatively about life is as just as dangerous now as it always has been.

Existential thinkers are hard to categorise. Their subject matter is life and existence although in many ways it appears to be the opposite as they often refer to death, anxiety and doubt, which appear to be the antithesis of life.

In undertaking the diploma I discovered a new word that has proved to be the cornerstone of a new way of thinking about life and all living organisms. That word is 'ontology'. Ontology comes from the Greek for 'being'. It is the name given to the philosophical study of the nature of *being, existence*, or *reality*. For me, it simply is the term that expresses the space in which free will is exercised in the pursuit of needs, goals and intentions.

Intention is another word that has a special place in the heart of existentialists.

Franz Brentano (1836-1917) argued that intentionality is the defining characteristic of human consciousness. In explaining this, Brentano removed the Cartesian distinction between mind and body. In his view there is no room between you and the objects that you relate to. Your 'mind', body and the world function in conjunction.

As we shall see in subsequent chapters the removal of the distinction between mind and body launched a truly Copernican Revolution.

Two of the main contributors to existential thought, Edmund Husserl (1859-1938) and Martin Heidegger (1889-1976) were deeply influenced by Brentano. Interestingly, Husserl was originally a mathematician and it seems that, yet again, mathematicians have much to tell us about life! He founded the philosophy and method of phenomenology as an attempt to understand 'reality' by setting aside (bracketing to Husserl's mathematical mind) all of our assumptions and interpretations of what we perceive. Heidegger was one of his pupils and was clear that human beings can only be understood if we are willing to abandon our certainty about the person or the ego.

Professor Emmy Van Deurzen headed the London School of Psychotherapy and Counselling when I started the diploma.

If we want to fully understand human existence we should not limit ourselves to the study of psychology. Emmy Van Deurzen

In the following paragraphs I will give a précis of Professor Van Deurzen's beautiful way of describing her views as gleaned from a number existential philosophers and writers.

Professor Van Deurzen is clear that, if we are ever to fully understand ourselves, we need to examine human living as expressed in our relationships to the physical world, to other people, to ourselves and to a network of meaning. As human beings we are complex bio-socio-psycho-spiritual organisms joined to the world around us in everything we are and do. In order to survive we need to be constantly connected, filled and fuelled. We need to contact and exchange with other creatures and with the systems that have been created for our more efficient survival.

The existential view is that we do not ever stand in isolation and a person cannot be an island. We are never but an element within a wider context and a thread within the tapestry that transcends us. We are always in relation, always in context, always connected to what is around us, always defined by what we associate with. Relationship is essential to our very survival and inspires everything that we are and do.

The quality of our engagement with the world is paramount. Therefore, we need not so much to pay careful attention to **what** we are, rather we should focus on **how** we are and to how we reshape and form ourselves by connecting and disconnecting with the context in which we find ourselves.

One of the fundamental premises of an existential approach is this awareness of people's contextual and relationship quality. It accepts the prime importance of our connectivity and constant change. This is of greater significance than at first sight may be obvious. As Professor Van Deurzen describes in her book *Everyday Mysteries*, it means that people are not just selves that go on to form relationships. Relatedness is rather the primary factor and the formation of the self is the result. The person is not viewed as an essence but as a medium through which life manifests itself.

Existential thinkers bring us all, especially those with big egos, down with a bump. Although we like to think of ourselves as independent agents, autonomous and constructing our own lives, the reality is that we are, in fact, essentially interdependent and contingent.

Building on the work of the philosophers, Professor Van Deurzen distinguishes four dimensions within which we have a different quality of relatedness:

- The physical, natural, material dimension,
- The social, public, cultural dimension,
- The personal, private, psychological dimension,
- The spiritual and ideological dimension.

The four dimensions are each spanned and dominated by a range of polarities. The contradictions and paradoxes that necessarily exist on each level are confounding but they are the vital source of our energy. As we discovered in Chapter 1, these polarities are like the positive and negative poles that generate the current of life. Life doesn't make sense without death and in the same way love doesn't make sense without hate.

Each of these four dimensions is interwoven with the others. Our journey encompasses all these levels but may emphasise one dimension at particular stages.

Physical Dimension

In the physical dimension we are bodies interacting within the physical environment. Basic motivating principles are those of survival and reproduction. We interact with the world in this dimension through our basic sensory and motor systems. Our body is the point of contact and action is our outlet. We achieve meaning through a sense of efficacy. The basic polarities that exist within this dimension are life/death and pleasure/pain.

Social Dimension

In the social dimension we are selves interacting within the world of other people engaged in contact either through cooperation or control. We achieve meaning through striving with others for the establishment of value. The basic polarities that exist within this dimension are love/hate and belonging/isolation.

Psychological Dimension

In the personal or private dimension we connect through are 'I' or 'self' to the internal world that we construct out of the experiences of the two other dimensions. Here we are concerned with creating the centredness that gives us a sense of stability, integration and selfhood. We create an inner sense of individuality. Meaning is created through a sense of self-worth. The basic polarities that exist within this dimension are identity/freedom and integrity/disintegration.

Spiritual Dimension

In this dimension we connect to the absolute world of ideas and the connection to a wider network, a sense of belonging to the scheme of things. The basic polarities that exist within this dimension are good/evil and purpose/futility.

In seeing the person as an element within a wider context and a thread within a tapestry that transcends us it is easy to jump to the conclusion that existentialists must be determinists. That is to say, they must simply see the individual as being like a reed carried along in a fast flowing stream. How can the individual be a force shaping his or her own life when so contingent, so interdependent, so built into a fabric of life? However, that is not the conclusion that existentialists tend to draw. Existential thinkers have *authenticity* as a key idea that provides the counterbalance to there being a sense of determinism and fate. In reading the disparate views from many philosophers and writers, it became apparent to me that this rather elusive concept encapsulates the following points. Whilst recognizing that we are contingent and codependent people who strive to be authentic realise that, through the decisions that they make, they are the architects of their own lives. As a consequence, they adopt powerful positions that enable them to:

- Take responsibility for their situation and feelings (they don't blame others for where they are or how they are currently feeling),
- Be aware of their own standards and values and live according to these,
- Always have a choice (even a choice not to decide).

What was the impact on me?

As part of the diploma course I had to undertake my own therapy sessions. The lack of trained existential therapists near to where I worked or lived meant that I had to find a therapist from a different branch of psychotherapy. I was recommended a local therapist that worked in a psychodynamic way and started weekly sessions.

Initially I started by lying on a couch, so you can see immediately the influence of Freud. I talked about what was going on in my life and in moments of 'self-flagellation' and self-blame I gained comfort from the voice behind my head.

Despite the investment of time and money I failed to be a good client. I had read enough about this form of therapy to know about 'transference'. So with regard to the hostility that I must have towards significant others I was aware that I should have been transferring this to my therapist. Yet, I could not do this. The therapist never stood in place of my mother/father/boss/wife. As I felt that the process was not working as it should I doubled my efforts by coming twice weekly.

I eventually had the courage to stop all together.

The process was fascinating. I did learn a great deal about myself. However, I also found out that self-knowledge and doing anything with it are two completely separate things. Real change, the sort that makes you behave differently, was incredibly difficult. I also know that the process of therapy can reinforce the stories that we tell ourselves. In the one-to-one therapeutic relationship there is no one there to provide an alternative account and a tentative hypothesis, through repetition, becomes the way that it is.

I am not against psychotherapy. There is a dire need for emotional and personal support and a process for gaining clarity. But in my case, I think that my mother's advice (you see, here she is) would have been better heeded, that is 'if you want a scab to heal, stop picking at it!'

Undertaking the diploma course provided a great opportunity for personal learning and reflection. During this I developed my own vision of purgatory. For me this was about sitting in a room with all the people that I have loved and having them watch a moment-by-moment video of my life including my internal verbal commentary that was up until this moment hidden from others. It is not the tedium that I would subject them to that I worry about. The process would reveal what I have hidden; the inconsistency and the lies. Most often the lies were to protect the feelings of others but sometimes they have been huge and convoluted to protect myself from being exposed as less than I would like to be.

There is a simple answer to address the worry associated with my particular view of purgatory. That is, not to say and do stuff that I will be sorry for everyone to hear and see in this life or an afterlife.

If only it was that easy.

Chapter 7
Bringing It All Together

In August 2007, I set up my own business and by September of the same year, i.e. one month later, the credit crunch looked like it was going to break the form of liberal capitalism that had developed over the last three decades. The financial meltdown spread into the wider economy and everyone battened down the hatches.

The world is now locked into the most prolonged economic downturn for eighty years. Britain, along with much of the rich world, is facing an apparently intractable slump. More than one million young people under the age of twenty five are unemployed. Living standards for those on low and middle incomes are on a decade-long decline and are now unlikely to return to pre-recession levels for another eight years.

Britain, it is argued, needs a sharp dose of austerity to get itself out of this mess. We have over-consumed, the country has run out of money and we now need to pay the price. Yet, while ordinary households are being squeezed, other parts of the economy are awash with money. Britain's top super rich continue to own more and more of the total wealth.

The combination of a continual squeeze on the masses and the accumulation of wealth in the hands of an elite has not proved a good recipe of success for the struggling small and medium enterprise – such as mine. However, I have not given up because I believe that I have been provided, by good fortune, the keys to the executive washroom. These are the in the form of a range of big ideas that bring together, into a coherent whole, many of the ideas that I have covered so far.

Big Ideas
The ideas that help integrate all the others encountered cover the nature of work, individual capability and 'flow'.

In his seminal work, 'Flow: The Psychology of Optimal Experience', Mihaly Csíkszentmihályi outlines his theory that people are most happy when they are in a state of flow; a state of concentration or complete absorption with the activity at hand and the situation. The idea of flow is identical to the feeling of being in the zone or in the groove. The flow state is an optimal state of intrinsic motivation where the person is fully immersed in what he or she is doing. This is a feeling everyone has at times, characterised by a feeling of great freedom, enjoyment, fulfilment and skill.

People are likely to feel 'in flow' when the challenge of the job is matched by their ability to get their head around it. But what does 'getting your head around the challenge' actually mean? Two people, Elliott Jaques and Gillian Stamp, both associated with the Brunel Institute of Organisational and Social Science (Bioss), provide the answer.

Jaques was considered within academia to be a heretic. He proposed that individuals tend to mature at given rates depending on their starting position. Ideas on the fixed maturation of human capability directly opposed the prevailing Standard Social Science Model that has dominated intellectual life. As we have seen, any questioning of this model implied that you supported ideas of biological determinism with all its evil implications. As such, much like Chomsky, views on Jaques were polarised. Whilst some loved his ideas, most hated them and in some academic circles, Jaques' name was besmirched by innuendo of him being some sort of Nazi.

Gillian Stamp worked with Elliott Jaques in the same institute and is now the head of the Bioss Foundation.

To understand what 'getting your head around the challenge' means we first have to understand 'work' and 'time'.

Jaques puts forward a very different view of work. Mechanical work can be clearly defined and is the execution of force to move an object in a given direction. But, what of the work of living organisms? In his studies at Glacier Metals, Jaques discovered that 'real work' there, for production workers and managers alike, was not the use of physical effort as in moving something but in the 'use of my head' in the sense of using judgement and making decisions.

Jaques coined the term 'organical'. Organical work is the exercise of living energy in making judgements, choices and decisions in order to achieve a goal in a given time. He also uses the term *locomoting*. Living organisms all have a future towards which they are 'locomoting'. Locomoting is the essence of being alive.

The planned future is in the very essence of locomoting. But let's not get confused. The future is not out there coming towards us. Jaques was clear that we are not travelling towards the future out there. Jaques addresses the nature of time and, in my estimation, the biggest insight that Jaques provides is his contention that there are two axes of time. He is clear that time is a human construct for organizing experience and, contrary to our understanding, does not go in any direction. Events proceed in a given direction and take a given time which humans learned to measure but time itself is not moving anywhere. The time, which is flowing from the past to the present to the future; all of which exists right now in the present, Jaques calls the time axis of intention.

Time in the science of human behaviour lies in establishing two dimensions of time. Firstly, how long something actually took to complete. This is the time of earlier and later. Secondly, how long something is planned to take. This is the time of past present and future. Elliott Jaques

I talk more about Jaques' concept of time in my book *It's About Time*.
A 5D world (3 spatial plus two time dimensions) really is necessary for locating a person engaged in work. Understanding the 5D world leads to the ability to measure prime aspects of human behaviour which have until now been taken for granted as non-measurable.
According to Jaques, an assignment is not just what an individual has been directed to do, it is a 'what-by-when' task, which is about 'time-span,' a simple, direct, objective measure of the size of a role. If two roles have the same time-span, they are the same size, regardless of what is involved in completing them.
This simple measure of a role's size has led to a connected measure of the size of the individual (i.e., the largest role an individual has the potential to carry out in work he or she values and for which he or she has gained the necessary knowledge and experience). What Jaques calls the person's 'time-horizon' is the longest time forward that an individual can plan and execute an assignment or reach a goal. Thus, a person can be identified as having, for example, a six-month, two-year or seven-year time-horizon.
Gillian Stamp broadens out the time-horizon aspect within Jaques' definition and refers to *perspective*. This provides an alternative to the commonly used notion of intelligence. Although psychologists and educationalists identify different forms of intelligence including emotional and physical, most of us think of intelligence as something that can be measured by IQ tests. Despite the fact that IQ has been bandied around for many years, in reality the only definition that can be properly identified for the concept is that it is the property that IQ tests measure! Perspective is just as easy to quantify as a score on a test and has the advantage of not being a circular argument!
We each have our own way of seeing, of reaching out into life and making sense of it. Perspective is not just about gathering facts. It's about how we figure out what to do when we can't find the facts to help us make a decision, or when facts contradict each other or, more painfully, when others contradict us. Perspective informs the ways in which we connect with the world and the people around us.
Perspective describes the 'focal width' of a person's worldview. It embraces the time-horizon that they use to make sensible decisions. For some people, their perspective is only about what is happening now - to them. For others, their perspective takes into account a broader view and entails thinking about historical antecedents and

43

long-term consequences.

Our perspective broadens as we learn from experience. As we mature our perspective of the world changes. Black and white becomes shades of grey. What was certain becomes nuanced. Where no meaning could be found, wisdom is revealed. This broadening is driven by curiosity. For some, their curiosity remains with the familiar - the tried and tested. For others, their curiosity leads them to go beyond accumulating experiences to link and make different connections. Our curiosity compels us to reach out and we perceive different connections, potentials, links, variables, possibilities; the raw material for making sense of the world. Some of us seek every possible detail on a subject and enjoy the accumulation of facts. Others seek for the connections and principles that link a number of situations or events together. Others revel in seeking patterns where none are obvious and working out the rules that govern them. Most of us deliberately seek to widen our perspective by reading odd and seemingly irrelevant material, watching interesting programmes, walking down unfamiliar streets, or putting ourselves in new or challenging situations.

When our breadth of thinking expands we can delay gratification and put in place plans that deliver benefits over a much longer time frame. And that brings us back to flow (a doe, a female deer!). People seek 'flow' because it is a reward in itself. The feeling of being in flow gives us energy and confidence, which feed accomplishment, which in turn boosts energy in a cycle of positive reinforcement.

Flow is not a luxury, but the essence of life. It inspires us to grow as we seek the pleasure of being in flow as often as possible. Yet it is often a questing beast that demands bigger and bigger challenges that can bring with it frustrations. It often feels as if our growing perspective has a life of its own as it seeks ever-farther horizons.

By seeking out associates and activities that challenge us in a positive sense – neither so much that we despair, nor so little that we doze off – we pave the way for a fulfilling, dynamic life, with opportunities each day to give the best in us and receive the best in return.

What was the impact on me?

The ideas provided by Elliott Jaques, as expanded on by Gillian Stamp and brought together in the concept of flow have made my life richer and more fulfilled. On the other hand, I am economically poorer and live with the real concern that I have entered a personal space that only I inhabit.

Before I found out about the nature of time, work and individual capability, I was doing quite nicely having climbed the greasy career pole to a reasonable height. I was a senior manager in a blue-chip organization and a Chartered Fellow of the Chartered Institute of

44

Personnel and Development (CIPD). Thereafter, it seems that I have slipped inexorably down. Now, I will have to suck on the bones of knowledge to find nourishment. After constantly trying to get the CIPD to provide a window for these ideas and failing, I have resigned from the Institute.

The business has not grown in the way that the strategic plan determined it should! The gamble hasn't paid off (yet) and in consolation I try to remind myself that it is only money that I have lost. But then I realise that it is only people with lots of money that can say, 'it's only money'.

What I have gained is a view of people that can be expressed as:

People are purposeful (intentional beings). We work towards goals within our particular mental-model guided by incentives that have a specific value to us. This mental-model is a way of making sense of the world and to anticipate (predict) the future. Anticipation guides our actions.

In other words people are active decision makers trying to get what they want in as sophisticated a way as their mental model allows.

This understanding is transformational.

Chapter 8
Hubris Country

I had a good idea, it was scalable and it had impact. These were, I was assured, the prerequisites of business success. I was confident that, despite the freezing economic winds blowing at the time, I would create the new order in Human Resource Management. Ambitious, possibly? Deluded, most definitely.

Success might have been assured, but what was success and how would I know that I had it once I had achieved it? These were questions that I couldn't easily answer. However, as a man that likes the good things in life, success in material terms was going to be a nice bonus.

Big Ideas

Having travelled quite a bit through airport lounges I decided to read some of the tomes I had gathered en route or at the least the ones on how to get rich quick and the many self-help 'success' recipes inspired by the original work of Napoleon Hill.

Everything your mind can conceive, you can achieve. Napoleon Hill

Hill's book, '*Think and Grow Rich*' sold 30 million copies. It is the sixth best selling business books of all time and is still in popular demand. The author pioneered the self-help industry. From Norman Vincent Peale, Tony Robbins, The Secret to Paul McKenna; all have been inspired by his fundamental concept.

Hill, struggling to escape an impoverished background in the Blue Ridge Mountains of Virginia, entered the world of journalism and gained access to steel magnate, Andrew Carnegie, at the time the wealthiest person in the world. Hill found out that Carnegie believed that there is a simple formula of success and was intrigued by this.

Impressed by his enthusiasm, Carnegie commissioned Hill to interview over 500 successful men and women including John Rockefeller, Theodore Roosevelt, Thomas Edison, Henry Ford, FW Woolworth and William Wrigley; all giants of America's political and business life. Hill then devoted the next 20 years of his life to identifying the laws of success. He planned to open the world's first University of Success but the Great Depression thwarted his plans and he began his masterwork, *Think and Grow Rich*.

Given the massive impact of the book, what of the man himself? Did he follow his own advice and become a great financial success?

Hill's own life is not one of consistent and sustained success. He abandoned his first wife because of lack of money despite his wife's family having bailed him out of a business failure. Hill went on to marry a younger woman, Rosa Lee Beeland, but she later divorced him, leaving him almost penniless. Ironically, Rosa later produced her own self-help book, 'How to Attract Men and Money'.

In a moment of rare humility, Hill observed, 'I had spent the better portion of my life chasing a rainbow. I had begun to place myself in the category of the charlatans who offers a remedy of failure which they, themselves, cannot successfully apply'.

Have you heard of The Secret? If you have, you may recognise it as Think and Grow Rich on steroids. The 'secret' is the Law of Attraction, 'You become or attract what you think about the most'. This isn't just an idea, it is a literal truth; it's a law. And it works every time No exceptions. At $35, the DVD, 'The Secret' by Rhonda Bryne and other gurus has sold over 3 million copies with the simple message; 'what we truly believe in our hearts and minds will come to us, good or bad'.

Of course how we think is important and some people do reap what they sow. Those who do the hard work often do see the benefits and people that do dumb things receive the penalty but The Law of Attraction offers something more; mysterious forces in the cosmos that can unleash unimaginable wealth, happiness and success.

Think and Grow Rich, The Secret and the vast library of self-help books can be summarized as 'You can do anything'. This is the basic American credo that has spread its optimistic veil of much of the rest of the world. It explains why so many success figures seem so prone to self-inflation.

If we look more closely, what are the implications of 'I can do anything'? Does it mean 'I can have everything that I want'? Does it mean, 'I can do whatever it takes to get what I want at whatever cost to anyone else'? Does it mean, I can 'rise above my character defects and become an attractive person, capable of leading others to the place that I want to go and the future that I see'?

We don't have to look far back in history to see the terrible results of people that really do believe that they can do anything. Surely there must be a countervailing pressure in our culture to counteract the worst excesses of 'I can do anything'? At first sight it looks like humanistic psychology including the writings of Abraham Maslow (1908-1970) provides the balance.

Lesson 1.01 on any Management Psychology course is likely to include Maslow's 'hierarchy of needs'. Maslow wrote that 'when people have satisfied 'maintenance needs' such as food, shelter and clothing, they progress to a higher level of gratification'. In his book, The Furthest Reaches of Human Nature, Maslow describes the path that he believes humans are on.

'It is true that lower-need gratifications can be bought with money, but when these are already fulfilled then people are motivated only by higher kinds of 'pay', e.g. belongingness, affection, dignity, respect, appreciation, honour as well as opportunities for self-actualisation and the fostering of higher values; truth, beauty, efficiency, excellence, justice, perfection, order, lawfulness, etc'. Abraham Maslow

The hierarchy of needs is a positive story about mankind naturally being on a higher path than simply acquiring possessions through trade or appropriating them by force. Yet, this widely accepted management idea fits nicely with the message, 'you can have whatever you want'. That is because the road to the higher life is through the first stage of acquiring stuff. Maslow doesn't say what enough is and leaves it to the positive power in his message that people will know when enough is enough. But that is not the evidence of history. What was a luxury for one generation becomes the necessity for the next.

By painting a rosy picture of what they consider to be human nature the humanists lead us into a false sense of security. They promote the idea that mankind is naturally on a journey to self-actualization, without fully explaining what this actually is. It suggests that once our leaders have got over the initial need to have security they will tend towards the higher aspects and through their own searching will lead us all to a better life. Some successful people do end up being role models and philanthropists but most do not. Human nature is not leading us anywhere.

Religion used to be the countervailing pressure to an impulse to acquire but in many ways the positive power orthodoxy has hijacked this. The new message is that 'God wants to prosper you'.

The 'I can do anything' message is loud and clear and perpetuates a series of myths:

Myth 1
Success is absolute and final. We will know it when we have it and that will be it; our desires will be forever satisfied. But then it turns out that wealth and power don't guarantee eternal bliss or shield us from all life's blows.

Myth 2
Money is central to success. Having lots of money frees you from the anxieties that those striving for a daily crust have to endure. Money though brings its own worries. Having gained it is one thing; losing it is entirely another. Losing is a pain we don't want to feel and we do crazy things to avoid. Money is power and power magnifies our faults.

Myth 3
Success will set you free. Success is measured by whether you have liberated yourself from the repetitive and humdrum. Yet, executives in high-powered jobs can feel imprisoned.

These myths underpin our consumer society. Consumerism has little to do with the factors that determine genuine well-being. We think our life will be much happier if we have a much bigger salary, house, car and more and better holidays. We over-estimate the impact of having nice things. Nice things do make us feel better but only for a short while. We often compare our future happiness against our current happiness and equate quantity with quality. When we amass the worldly goods that we aspired to own the comparisons we make then become relative to others. We want more than the person next door or the CEO of the rival company. Success, like all things, is relative. I am also reminded of Gore Vidal's maxim, 'It is not enough to succeed; others must fail'.

The simple message of 'you can do anything' and the success myths have given rise to many examples of hubris over the last decade. Hubris can be understood as the ego becoming swollen with success and filled with the sense that I can bend the world to my will. It is a psychological blindness. Signs of this blindness are there in many of the so-called Masters of the Universe. These signs are a tendency to ignore or discount negative information about themselves or their business, or an inability to perceive the needs of others and behave as if they only existed to serve their needs.

Paradoxically, hubris seems to flourish in conditions of greatest uncertainty. Fuelled by notions of power and money, hubris is rife in politics, academia, the legal profession and the finance industry. Just when you have the greatest need to doubt, hubris and the focus on our own special gifts cuts through and says 'my way or the highway'.

It is clear that a success culture provides many traps. One to avoid is the passion trap.

The message is relentless and seductive. That is, 'live your passion'. Passion of course matters. Enthusiastic commitment to usefully deploy our fundamental talents and skills is more rewarding than the dull lethargy of half-hearted effort, but finding a passion that sings to the soul can be like finding the Holy Grail.

There is a pressure to find this passion and special purpose. However, the more emphasis that we place on being fulfilled and happy the more miserable we become as it becomes evident that we don't love every minute of the work that we have.

The 'discover your passion' movement has inadvertently led to a great deal of misery and heartache as people risk all, including foregoing a steady income and family life, through ventures that seek the Holy Grail but rarely pay off.

What was the impact on me?

Having read and inwardly digested the *Think and Grow Rich* mantra, I plunged into turning my business idea into reality. This involved the

50

simple notion of matching people to jobs by ensuring that jobs and CVs were written using a common language and similar dimensions. Using similar Internet-based templates for job descriptions and capability-based profiles meant that shortlist selection decisions could be automated.

I followed the *Think and Grow Rich* formula most of the time with the exception of having a monetary or lifestyle goal in mind. Whilst comfort and security appealed, I genuinely did not aspire to wealth beyond the means of avarice and I was only keen on keeping what I had gained to date. Money was of interest but my main motivation was to liberate a set of ideas that had the power to make a difference. Nevertheless, despite the slight change in recipe, I was still convinced that it was just a question of time before success was achieved.

It still is a question of time! I am still waiting. I can take some comfort in also knowing that other simple ideas took a long time to actually be implemented including the ubiquitous ballpoint pen and zip.

What the adherents to Hill's ideas will say is that I didn't follow the recipe exactly. No wonder that the cake hasn't risen.

The lesson from Hill's life and work to me is; think very carefully before chasing a rainbow. Better than chasing it yourself, package it up and sell it to others. Then you'll find a pot of gold. This is the business model of the self-improvement industry.

What they don't tell you in *Think and Grow Rich* and its hundreds of variations is that setting up a business that intends to offer something very different in the market will take you on an emotional roller-coaster. You will experience a range of feelings from near euphoria when someone says that they like what they have seen to desperation when it seems that no one is listening including partners and bankers. The books don't tell you about the constant, almost 24/7, thinking about the idea and how to make it live and the incessant rewriting of the business plans and strategy papers. What they don't tell you is what you have to give up long before it is clear that the ideas will pay off. What they don't tell you is that despite the belief, the effort, the financial commitment and the heartache, people might not get it.

I got the future wrong because, for a while, I unquestioningly ingested a culture that reinforces flawed assumptions about the nature of uncertainty and the dynamics of happiness and success.

Chapter 9
The Power of Ideas

The previous chapters detail the big ideas that have influenced me and shaped my career. Before moving on to explain how I integrate these ideas into a new model of human capability, I want to explore the power of ideas and identify how these are the 'goggles' through which we see the world. The goggles of industrialised and mechanised mankind have proved to be useful in the past but they are also limiting our vision to a bigger picture.

Ideas are the lifeblood of humankind. We cannot live without ideas. We think with and through ideas. Upon them depends what we do.

In essence, life and living is the process of doing one thing instead of another. To help us choose what to do we use ideas. Ideas help us think and through them we make sense of the world, society and our own life.

That which we call thinking is generally the application of pre-existing ideas to a given situation or set of facts. Some of the ideas are ideas of value and we evaluate situations in the light of these value-ideas. As such, the way that we experience and interpret the world obviously depends very much on the ideas that fill our minds.

We often notice the existence of more or less fixed ideas in others. We call these prejudices when they appear to have seeped into a person without judgement being applied. Of course we are not like that at all. Or are we? Certainly when it comes to thinking about human nature there are many unhelpful prejudices.

The world of ideas has shaped our view of human nature and our views on human nature underpin everything that we do in the world of work and beyond.

Ideas about human nature at work are clouded by serious, invasive misconceptions that have continually fueled the development of ill-advised economic and managerial systems. Drawing on more than 100 years of thought, observation, analysis, and experimentation it is clear to me that these shortcomings are the result of using theories and models that do not embrace the whole person.

Understanding human nature is not simply a luxury of the intellect. It matters deeply.

Let's explore the ideas that are around now and how they shape our views of the world around us.

E F Schumacher enumerates five leading ideas all stemming from the 19th Century that have dominated our thinking and influenced our views of human nature, namely: Evolution, Class Struggle, the Dynamic Unconscious, Positivism, and Relativism,

Evolution

There is the idea of evolution; that is higher forms continually develop out of lower forms as a natural and automatic process. Evolution is the change in the inherited characteristics of biological populations over successive generations. Charles Darwin was the first to formulate a scientific argument for the theory of evolution by means of natural selection.

Class Struggle

There is the central idea within Marxist ideology that the higher manifestations of human life such as religion, philosophy, art etc. (what Marx calls 'the Phantasmagorias in the brains of men') are nothing but superstructures erected to disguise and promote economic interests, the whole of human history being the history of class struggles.

Class Struggle is the tension, or antagonism, which exists in society due to competing socio-economic interests and desires between people of different classes.

The Dynamic Unconscious

In competition with the Marxist interpretation is the Freudian interpretation that all higher manifestations are the result of the libido. Freud defined libido as the instinctive energy or force, contained in what he called the id, which is the largely unconscious structure of the psyche.

Freud viewed libido as passing through a series of developmental stages within the individual. Failure to adequately adapt to the demands of these different stages could result in libidinal energy becoming 'dammed up' or fixated in these stages producing certain pathological character traits in adulthood.

Positivism

There is the triumphant idea that valid knowledge can be attained only through the method of the natural sciences and hence that no knowledge is genuine unless it is based on generally observable facts.

Positivism assumes that there is valid knowledge only in scientific knowledge. Obtaining and *verifying* data that can be received from the senses is known as empirical evidence. Underlying positivism is a belief that society operates according to its own laws, much as the physical world operates according to gravity and other absolute laws of nature.

Relativism

There are many forms of relativism, which vary in their degree of controversy. The term often refers to *truth relativism*, which is the doctrine that there are no absolute truths, i.e., that truth is always relative to some particular frame of reference, such as a language or a culture. We have already seen how this idea set the frame for

54

Standard Social Science Model and how it precludes certain areas for debate. Anything outside of a Relativist paradigm must be biological determinism and that is what another age would have described as the Devil's work.

The essential character of all these big ideas emanating from the 19th century is their claim to universality. For instance, evolution takes everything into its stride, not only physical phenomena but also mental phenomena such as religion and language.

The power to uncover the mystery of human nature must be here in one of these ways of understanding the world?

So much progress has been achieved through the application of the scientific method; surely this is the way to enlightenment? A defining characteristic of human progress is our ever-increasing ability to make the uncertain certain. Our lives today are what they are, better or worse (largely better I suggest), because we have used the scientific method to tease apart underlying cause and effect relationships in the physical world. This has not only satisfied our curiosity, it has given us the power to predict and control outcomes that matter. As a consequence, much of what once appeared to be in the realm of the Gods has become the province of mere mortals.

Our collective belief is that anything beyond our grasp is only temporarily so and it is only a matter of time before this will fall under our dominion so long as we stay true to the scientific method and observe the processes and mechanisms on which life must be based. However, when we, that is, human beings, become the focus of the scientific method, rather than revealing the essence of being human, it seems that we are reduced to merely a collection of mechanisms that constrain rather than enable our freedom.

Well, if the scientific method has its shortcomings surely Freud's big idea must open the possibility of revealing the nature of mankind through discovering the unexplored continent of the dynamic unconscious. Yet, psychotherapeutic culture, and not just the adherents to Freud's psychoanalytic approach, tends to focus on the internal world of the psyche and its cognitive and emotional processes. Little attention is paid to the world in which the person lives. This is like trying to find out about what makes a tiger a tiger without fully understanding the environment in which tigers live their lives.

For all their power and promise these big ideas miss out the essence of humanity as they all intrinsically assert that what had previously been taken to be something of a higher order (such as Truth) is nothing but a more subtle manifestation of the lower. This process of reduction, inevitably leads to the idea that life is essentially futile.

The classical Christian culture of the late Middle Ages supplied

55

humankind with a complete and coherent system that provided a detailed picture of our place in the universe and provided for redemption and salvation. This system has shattered and has largely been replaced by a set of fragmented and sometimes contradictory ideas.

Almost to the other extreme of the classical Christian culture, many philosophers have replaced concepts of devotion, love, virtue and redemption with nihilistic ideas. Bertrand Russell for instance claimed that the scientific theories leading to the conclusion that the whole universe is simply the outcome of an accidental collocation of atoms, 'if not quite beyond dispute are yet so nearly certain, that no philosophy that rejects them can hope to stand..... Only on the firm foundation of unyielding despair can the soul's habitation henceforth be safely built'.

Although despair is an existential issue, the big ideas above do not directly address the existential concerns of life and human living as their primary focus. In fact, in many ways they presuppose that people can be treated as separate units, which can be examined, analysed, diagnosed and classified like mechanical objects. However, taking people as objects does not make sense; people are defined by their relationship to a physical world, to other people, to themselves and to a network of meaning. The reality is that people as organisms exist only in relation to a context and an environment.

Not only is there a process of reduction going on which fails to see human being as essentially in context – at the heart of our understanding of ourselves is a blind spot. This precludes us from seeing the full process of social reality formation. In everyday experience we do not see the coming-into-being of social action; we do not see its descending movement from thought and consciousness to language, behaviour and action. We see what we do. However, we are usually unaware of the place from which we operate when we act.

It is difficult for us to become aware of the place from which we operate because the ideas that we use are often the instruments and not the results of our thinking. This is similar to the way in which we see. Just as we can see what is in front of us, we cannot easily see that with which we see, namely the eye itself.

The blind spot leads us to think we know about what makes us behave in the way that we do but we don't really and this has serious repercussions. We form theories about how and why we do things but these are often misguided post-rationalisations. These include misunderstandings about our thinking process. If thinking is considered in management science it is normally limited to the classical aptitudes of logic and reasoning. But thinking is much broader than that and thinking about thinking defines us as human beings. Thinking about thinking moves us closer to understanding the

56

essence of being human and to that which we term human nature.

What is lacking and not just in management science, is a unifying idea about human nature.

The true nature of a person is left out in most psychological theories. The gap is filled by religious or quasi-religion ideas, which include the positive humanistic notion that humans are on a path to self-actualisation. This largely ignores the fact that we are also on a path to environmental catastrophe. If psychology falls short, the theories in economics are openly partial and selective. Here people are cast as either producers (costs) or as consumers (rational beings making informed choices).

Can we look to education to help us find the whole person?

Formal education is the greatest resource of modern times and is increasingly available to the world's population. Education is underpinned by a value system and that system places 'convergence' on the apex of the value pyramid.

By convergence I mean the ability to give the 'correct' answer to standard questions that require coming up with the single, well-established answer to a problem through logical reasoning. It involves deriving the single best, or most often correct answer to a question. Convergent thinking emphasizes speed, accuracy and logic and focuses on recognizing what is familiar, reapplying known techniques and accumulating stored information.

Convergence (with its opposite – divergence) has proved to be a most useful construct.

Convergent problems do not as such exist in reality but are created by a process of abstraction. When they have been solved, the solution can be written down and passed on to others who can reproduce it without the mental effort that is required to find it.

The physical sciences and mathematics are mostly concerned with convergent problems. That is why they can progress cumulatively and each new generation can begin just where their forebears left off.

If human relations could be reduced to convergent solutions in family life, economics, politics, education and so forth there would be no more human relations but only mechanical reactions. Thankfully life is not like that. Not everything can be 'mechanically' solved by the application of known principles and experience. Life throws us challenges that can only be solved divergently. We have to forge new and surprising connections; we have to pretend and ask 'what-if' and we have to picture and imagine a different future.

Life forces us, as individuals, to strain to a level above ourselves. They demand forces from a higher level and this brings love, goodness, beauty and truth into our lives. It is only with the help of these higher forces that the opposites can be reconciled in the living situation.

So, let's now move on to look at a new model of human capability. In

57

order to fully appreciate this, I would ask you to be as aware, as much as you can, of the goggles that you typically use to view and understand the world. The new model will ask you to broaden your perspective and it will question the constructs and polarities that are embedded in our world-view such as that of 'mind-body'.

If you can take on a new perspective you will have a better view of the whole human being. When the whole person is taken into account there is a very different future to be imagined.

Part Two

The New Model of Human Capability

Chapter 10
The New Model

To a larger or lesser extent I have introduced you to the people and their ideas that have influenced me. That is, with the exception of one person, Daniel Kahneman. Kahneman is a Senior Scholar at Princeton University and Emeritus Professor of Public Affairs, Woodrow Wilson School of Public and International Affairs. He was awarded the Nobel Prize in Economics in 2002.

Kahneman, together with his colleague Amos Tversky published an influential paper in 1974 regarding judgement in conditions of uncertainty. Kahneman and Tversky showed that people routinely assess the probability of an uncertain event or the value of an uncertain quantity by relying on rules of thumb that reduce complex calculations to simpler judgemental operations. This was the first of many insights on how we handle uncertainty and, in the process, how we manage to trick ourselves.

In 1979 Kahneman and Tversky published a paper in *Econometrica* entitled *Prospect Theory; An Analysis of Decision Under Risk*. In this they categorically showed how people behave differently when confronted with gains and losses. What they found was counter intuitive. People were more likely to take risks to avoid losses than they were to achieve gains. That is, we find it much harder to stomach a loss than to miss out on the chance of gain. People go out more on a limb to protect what they have, than gain more.

Some of the most well known events in finance followed exactly this behaviour. Nick Leeson, faced with huge losses from a gamble gone wrong, gambled yet more in the hope of reversing his losses. Instead he lost more and eventually broke Barings Bank. The same motive and loss aversion behaviour was true in the case of Kweku Adoboli who lost £1.5billion of UBS's money. Instead of recognizing this as normal behaviour, given the system in operation, these two people in particular were labelled rogue traders and sentenced to lengthy jail terms.

In 2011 Kahneman published an eminently readable book, *Thinking, Fast and Slow* that tells more of the systematic biases that people use to turn uncertainty into certainty.

The ideas that I have detailed throughout the book, together with insights that Kahneman provides from cognitive psychology, form the basis of the new model of human capability detailed below.

One System
The absolute foundation of the model is the principle that individual

behaviour is a function of the whole individual organism as a total information processing system.

Before moving on to explain the capacities (enablers) and capabilities (outputs) of the new model, I don't want you to think that the information processing system that I am about to describe can be likened to that of a telephone exchange, super computer or the Internet.

When talking about human beings it is easy to reach for analogies. However, when we draw upon the most complex things humans have made to date it is easy to jump to the conclusion that, as these are the outputs of the human brain, they must in some way reflect the inventor. Analogies with telephone exchanges, super computers or any human invention might be helpful but they are also misleading and take us straight down two dead-ends that many philosophers and scientists have taken and still take today. One of these dead ends is the attempt to locate the centre of reasoning. The starting point for this new model of human capability is that all cognition is embodied; that means that we make decisions with our whole body and not only with our brain. Attempts to find the seat of thinking or the essence of the human mind or the locus of control are doomed to failure. Living is locomoting and locomoting is a whole body process.

The other dead-end is thinking of the brain as some sort of electrical device that simply responds. **All living organisms are first and foremost intentional. They reach out to the world and beat it to the punch.**

Let's turn to the capacities and capabilities, modes and selves within the new model. However a brief caveat; if it is not evident from the earlier point, this is descriptive language and it should not be taken that I am identifying separate processes taking place in separate parts of the brain.

One System, Five Capacities

All humans have interdependent capacities that, together, build and maintain a model of their personal world, which enable judgement and goal-directed behaviour. These capacities are:

- Powerful categorisation,
- Cause and effect reality testing,
- Associative-activation,
- Engaged awareness,
- Disengaged awareness.

Categorisation

All perception involves categorising. If you see something that you have never seen before you will already have categorised it as 'something that I have never seen before' even before you utter the words to yourself. At birth, and before, the nervous system distinguishes between light and sound stimuli and deals with the categorised information very differently. At the other extreme, in complex social arrangements sophisticated categorising is very evident. A person reacts to others depending upon on how he or she has categorised them. Probably the most general dichotomous category is 'Us versus Them'. Categories do not have to be verbally expressed and, as such, all living organisms are categorizing systems to a larger or lesser extent

Within humans the categorising system is highly sophisticated and is structured according to a 'universal mentalese'; a set of rules that includes pattern recognition and detecting similarities. We are pattern seekers oriented to detect regularities and our mental software is set-up to generalize according to similarity. A child who continually echoes back a parent's sentence verbatim would not a called powerful learner; powerful learners generalize to sentences that are similar to their parents' not to those sentences exactly. Detecting similarity is the mainspring of learning. The meal we ate yesterday is not the same meal that we ate today but our use of the word 'meal' is an explicit recognition of similarity, some replication that helps us anticipate events. The foundation for our making sense of our world is this continual detection of repeated themes, the categorization of these and the use of them to segment our world and act within it. The themes we recognise can be concrete as in our noting of recurring 'toothbrushes' and 'taps' or they may be very abstract as in 'trustworthiness' and 'truth'.

The categories that a person utilises can be explored through the use of Repertory Grid Technique. This was created by George Kelly, whom I introduced in Chapter 1.

Cause and Effect Reality Testing

Linking cause and effect is not a one step process and does not involve the brain simply linking a series of stimulus and response chains through trial and error. First and foremost all living organisms are in the anticipation business. As such, linking cause and effect starts with a prediction such as situation A would lead to another situation, B. Whenever the prediction proves correct a conclusion is made that A had caused B. As such, a conclusion is drawn about cause and effect, not from observing a B preceded by an A, but through an expectation being realised.

Associative-activation

When processing any stimuli there is an automatic search for a connection.

Take the random words, 'knife' and 'blood'. There was no particular reason but you automatically assumed a temporal sequence and connection between the words forming a sketchy idea with a link in the chain possibly being 'accident' or 'fight'. You did not consciously think about it and you could not stop it. The event that took place on seeing the words is a process that Kahneman calls 'associative-activation'. Associative-activation happens in an instant. Again, it is not a process limited to that which can be verbally expressed.

Associative-activation involves a spreading cascade of activity as one stimulus triggers many others. This process is far from random and an essential feature of this complex set of events is its coherence. Each element is connected and supports and strengthens the others.

Associative-activation is the key process in building skills, knowledge and expertise. Links are established between circumstances, events, actions and outcomes that occur with some regularity either at the same time or within a relatively short time interval. As these links are formed and strengthened the pattern of associated ideas comes to represent what is known and familiar.

Engaged-Awareness

People (and all living organisms for that matter) have an in-the-moment capacity to attend to the world and orient to new situations and stimuli. A capacity for surprise is the essential aspect of life. Engaged-awareness registers interest and we orient to what is surprising. Surprise is the most sensitive indication of how we understand our world and what we expect from it.

Disengaged-Awareness

Fully developed adults have the capacity to disengage from the 'here and now' and to share with others their inner world through the power of language. Language changes the nature of the world that we live within. It provides the capability for self-talk which we take to be consciousness.

The central feature of this disengagement lies in the ability to articulate our knowledge of the world in more complex and general terms. It is general in the sense that it is not tied to ongoing actions in immediate time and place but can be used to represent classes of things and classes of actions at various times and related to various types of action – or in fact no action at all.

Generalised propositions, when externalised as statements, whether orally or in writing (words, drawings or designs), then have the truly extraordinary quality of taking on potentially permanent existence of their own – to be available to oneself and to others through time.

One System, Two Modes of Thinking

The capacities of categorisation, cause and effect reality testing, associative-activation, engaged-awareness and disengaged-awareness work as one system with reciprocal links enabling one capacity to influence others. This system provides for two distinct, yet interdependent, thinking capabilities.

Psychologists have been intensely interested for several decades in these two capabilities (which we can refer to as modes of thinking), namely effortless instant interpretations/predictions, which I will refer to here as Mode 1, and effortful thinking, which I will refer to as Mode 2.

Mode 1 - Effortless Instant Interpretations/Predictions

Mode 1 operates automatically and instantly with little or no effort and no sense of voluntary control. It has been shaped by evolution to provide a continuous assessment of the main problems that an organism must solve to survive. Mode 1 provides the answers (not necessarily verbalised) to, how are things now? Is there a threat or opportunity? Is everything familiar? Should I approach or avoid? Not only does it provide the answer it also calls for action.

Mode1 is naturally geared to discriminate sharply; it is hard wired to make snap judgements about the other's status and their potential threat. The continuous struggle for survival requires us to distinguish those that will protect us and those that might attack, exploit or undermine us. The fear of the stranger is hard-wired, as is the need for community.

Mode 1 is very efficient. Its interpretations, based on familiar situations, are often accurate and so too are its short-term predictions. The predictive capability is based on being able to identify similarities.

Mode 1 does have limitations including the fact that mere exposure to an object leads to a sense of familiarity and familiarity reduces engaged awareness. Thus, we can be lulled into a false sense of security. It also cannot be turned off. We are addicted to interpretation and prediction.

Instant interpretations/predictions are prone to a range of bias including:

- Answering easier questions than were asked by applying heuristics - rules of thumb,
- Seeing causality just as directly as we see colour and attributing causality even when it is not the case,
- Priming effects which influence our selection process,
- The initial impulse to believe,
- Pattern recognition when there is none,
- Certainty of one, and only one, interpretation.

These biases are fully explained in Daniel Kahneman's, *Thinking, Fast and Slow* and so I won't expand on them here.

In essence, humans have the ability to jump to conclusions. Often these are helpful. Sometimes they are very damaging especially when they are based on the predilection for causal thinking (even when there is no cause) and pattern recognition (when there is no pattern). Amongst other side effects, these biases expose us to serious mistakes in evaluating the randomness of truly random events.

Mode 2 - Effortful Thinking

Mode 2 is a reflective capability that allows us to monitor and control what arises automatically and can suppress or modify actions. It is the effortful mental activity that we carry out including logical analysis and complex computation.

Mode 2 undertakes detailed processing and involves following rules and comparing objects on several attributes.

The automatic operations of Mode 1 generate surprisingly complex patterns of ideas but it is only Mode 2 that can construct thoughts in an orderly series of steps.

A big distinction between the modes relates to the effort required. Mode 2 only operates when we pay attention. 'Pay' is a good word because it suggests that attention is limited and when we pay we become depleted. If we go beyond our budget we fail. When this happens people can become effectively blind even to stimuli that normally attract attention.

A general law of least effort applies to cognitive as well as physical exertion. The law asserts that if there are several ways of achieving the same goal people will eventually gravitate to the least demanding course of action. In the economy of action, effort, including cognitive effort, is only expended when the benefit outweighs the cost. Finding the easiest way of doing something is built deep into our system.

Two Capabilities, One Decision Making System

The operations of Mode 2 are often associated with the subjective experience of agency, choice and concentration. When we think of ourselves we identify with mode 2; the conscious reasoning self that makes choices and decides what to think about and what to do.

However, just because it is possible to identify a mode that provides instant interpretation and one that requires effort, it should not be assumed that there are two decision making processes. Mode 1 and 2 are part of the same decision making process; one that is not open to observation. In this respect our thinking processes mirror the rest of living organisms.

This is completely at odds with our sense of how we as individuals, as representatives of the rest of humanity, make decisions. There is an almost universal belief that the best decisions are rational decisions. Rational decisions are those that we believe we arrive at by reasoning with objective facts.

The sense of taking rational decisions is a trick we play on ourselves and I will cover this in greater detail in the next chapter.

66

In order to assess the breadth and depth of a person's capability to make sensible decisions it is necessary to evaluate their perspective, which, as we have seen in Chapter 7, includes the ability to handle varying time-horizons.

One System, Two Selves
The one unified system gives rise to two selves, 1) an experiencing-self and 2) a reflective-self.

The experiencing-self is in the moment of occurrence and comprises attention and the feelings that the situation generates. The experiencing-self attends to what interests us and whatever seems likely to serve our personal desires. It can be an inquisitive beast, keeping its nose down to the scents and sounds of the trail and be blind to its wider surroundings.

The reflective-self is the creator of the story that we tell ourselves. It is also the self that can widen the area of attention when the moment-by-moment impulses are held in check.

Our reflective capability and the story that we tell ourselves influences the experiencing-self through directing what we attend to. Similarly, what we attend to, in-the-moment, influences our reflective-self.

The experiencing and reflective-self can work in harmony through a process of focus and expansion. Alternatively, they can work against each other to create ceaseless chatter, which excludes any meaningful appreciation of the moment-by-moment experience.

Our reflective-self develops as we mature. Initially, children learn to discriminate others in very crude terms. Others are goodies or baddies, kind or wicked, givers or takers. Initially we tend to like those who are like us and dislike those that are different but gradually we come to see the virtues of complementarities. At first we are inclined to believe that good and bad are inherent characteristics and that once you can tell who is good you can trust them on all things forever. Eventually we come to realise that the world is not divided between good an evil in this manner and that all of us are good in some situations and bad in others.

Through the capacity for reflection, we fashion a sense of self that increasingly becomes more defined. At first this entity is probably nothing more than a vague sense of bodily-me in interaction with the physical world and then later a sense of social-self develops. It is only when our ability to reflect upon our experience increases sufficiently to form mental concepts that we begin to create a private space in our inner world. We learn to protect this new and fragile-me and get disturbed when social appraisals of us clash with a sense of our selves.

Our sense of self is more easily formed if we experience the world of others as a relatively safe place, which will not be crushed by others.

Where the physical and social dimension is experienced as unsafe, a prematurely defined self may emerge with very tightly drawn boundaries. If all goes well however, we create a realm of warm security for inner dialogue. Into this realm of the personal-self we allow certain loved and trusted people, special objects and animals.

The sense of self is often fragile and easily damaged and to protect it we employ a range of overt and covert defence mechanisms. Sometimes we make the wrong choice when we turn to authority figures to help protect us not only physically but also as psychological entities.

In our personal world it is our thoughts that guide us. Inner dialogue is a tremendous potential strength. However, our reflective-self is prone to errors and misinterpretations. Key misinterpretations of the reflective-self are:

- We are rational decision makers because we articulate the reasons for decisions,
- Time flies like an arrow- so we can predict what will be in the future,
- Our model is reality,
- Because we have a separate body, we have a separate mind too,
- We think that we know what will make us happy.

We will look further at these misinterpretations in the next chapter.

What has happened to emotions in all of this you may well ask? A charge that could be levelled against this model of human capability is that it is too 'mentalistic'; too geared towards thinking. However, we do not have to accept the cognition-emotion division that such criticism rests on. This division is jargon descendent of the ancient dualities of *reason* versus *passion*, *mind* versus *matter*, *flesh* versus *spirit*, which has lead to dualist psychologies.

Emotion is a 'hydraulic' concept, which makes for very great problems. Hydraulic concepts postulate some force (motive, instinct or drive) within ourselves impelling us to movement. In simple terms, hydraulic concepts involve a vision of some kind of ginger-pop fizzing about the human system. However, we can safely reject hydraulic theories that seek to explain the process of living. It is entirely unnecessary to account for movement if we begin by assuming the fundamental thing about life is that it is all about locomoting; it moves, it goes on. It isn't that something makes you go on; the going on is the thing itself. It isn't that motives or drives make us come alert and do things; our alertness is an aspect of our very *being*.

A distinction is only useful if we can do things with it. We may find that we can do more without the cognition-emotion distinction than we have been able to do with it. Understanding that the 'mind' is embodied and that reason and emotion are not separate can help us to see a fuller picture.

In order to avoid the cognition-emotion dualism, I am going with the suggestion made by Kelly, whom I introduced in Chapter 1, that we can focus our attention on certain specific constructs such as anxiety, hostility, guilt, threat, fear and aggression but define them all as 'states of awareness'. These feelings let us know that our mental models, i.e. the way that we construe and understand the world, are in transitional states.

Chapter 11
The Stories We Tell Ourselves

In the last chapter, I identified the key misinterpretations of the reflective-self. Here, I expand on these and the important knock-on effects.

We are rational decision makers

The sense of taking rational decisions is a trick that we play on ourselves. The organical work process involving real decision-making is totally ineffable. This means that it is incapable of being described. It is also not open to observation. You can observe the outcome of the process but you cannot directly observe the process itself. The inaccessibility of the process and the fact that the process is the same in all living organisms gives life its sense of mystery. The scientific community is not used to the idea of systems and process that are not directly observable. It is going to have to get used to the idea in this case.

Our reasoning capability helps us to focus our decision-making process. It is not in itself a decision-maker. What we do is use consciously available knowledge to focus our attention and let the ineffable choice process take its course. As Jaques put it so clearly, 'There is no provision for purely rational decisions for there is no such process'.

All acts of choosing and making judgements, in fact all free will related activities, take place behind a locked door.

Post-rationalizing is part of the human condition. If, for a moment, we could suspend the internal dialogue we would recognise that at a decision point we do not know what decision we have made or are going to make until after we have made it. Even that decision has not been finally made unless we have acted and committed resources to it and would cost us something to withdraw from it or change it.

In everyday experience we do not see what precedes action. We see what we do. We also form theories about how we do things. These theories tend to suggest that we act rationally, having fully comprehended the inputs and consequences.

We need to understand that when someone comes to a 'rational' decision, say to devote themselves to a new career, this is not the simple result of a wilful self making a deliberate, conscious choice to assert itself in this particular way. Neither is it sufficient to frame the decision as the outcome of an unconscious push of certain drives and determining factors from childhood that bring about the commitment. Every move we make, everything that we decide is the outcome of a process which is behind a locked door and involves a multitude of influences including elements of past history, present awareness and future expectations. There are influencing factors of education, class and culture. There are situational, and contextual factors. There are hormonal and genetic influences that affect the move we make. There is a range of internal conflicts and systematic bias that often cannot be verbalised. None of these factors alone determines what will happen, or rarely so. However, if we ask someone to explain their decision they will do so in a way that convinces them that their decision is a result of their well-thought-through rational considerations.

Time flies like an arrow

It is logical to describe life as series of moments and that the value of an episode (say a fortnight's holiday) could simply be the sum of its moments. But this is not how the human mind represents episodes. The reflective-self, as I have described it in the last chapter, is selective, tells stories and can be economical with the truth.

As master storytellers we focus on the beginning, the highlights and the end. That is, we pay attention to selected moments and neglect what happens at other times. As such, duration of experience is not a big factor for the reflective-self. We also make important what we think about (the 'focusing illusion' according to Kahneman) and what we know for ourselves is that we don't think so much about that which is familiar.

Our ability to retell the moments of our lives and recast lived experience in other ways is illustrated in the case of a partner who finds their other half has been cheating on them. The shared moments, whilst experienced at the time as happy, turn to dust and are seen in a completely different light.

It appears that we are good with stories but not the processing of time.

Not only don't we process time well, the story that we tell about time is completely fictional. Time is not running anywhere and it is not aimed at a specific point in the future. There are no inevitabilities. The future is now. The future is here, right now at this very moment in the form of a current goal alongside the present, which is our current outlook, and alongside the past, which is our current history.

Our model is reality

Humans can withstand lots of things but uncertainty isn't among them. Uncertainty keeps us in a state of anxiety and our mental software isn't well equipped to deal with it. Fortunately we have a process that quickly releases us from our discomfort. This process enables us to jump to conclusions.

In the last chapter I talked about two separate modes of thinking. An essential design feature of Mode 1 is that information that is not retrieved (even unconsciously) from memory might as well not exist. Mode 1 excels at constructing the best possible story that incorporates ideas currently activated. It does not and cannot allow for information that it does not have. The measure of success is the coherence that it manages to create.

Mode 1 is radically insensitive to both the quality and quantity of the information that gives rise to impressions and intuitions.

In addition to Mode 1, Mode 2 enables more thoughtful consideration. However, this requires effort and often the reward of checking the facts that Mode 1 uses to jump to conclusions is not worth the bother, so we let the story stand. Even if we make the effort, as we have used our rational qualities to come up with an 'objective' position, we believe the output and are blind to our own bias.

On the basis of our two Modes of thinking we can predict and anticipate events. However, this comes at a price. In preferring certainty to uncertainty we take the model to be real and concretise the world.

We have a mind that is separate from the body

Attempts to understand the behaviour of living organisms, and in particular the behaviour of human beings, have been plagued by the question of where this mysterious behaviour comes from. Is it a consequence of the actions of minds and bodies? If so, what is then the nature of the mind-body relationship? Or does it derive from brains, or other parts of the system, or from minds encapsulated with brains?

It has long seemed useful, and well before Descartes, to worry about the relation between mind and matter and to talk about human beings in terms of body and mind or of body and soul or mind, body and soul. The idea of a body has seemed real enough because we can see what it is doing and where it is going and we can see the muscular system at work moving it. As we observe parts of the body at work we see that the eyes are looking, the ears hearing, the nose smelling and so on. However, once we separate out the sense organs and the organs of perception, the problem becomes one of getting a usable formulation of who is doing the seeing, hearing and smelling. More sharply, and in this instance, who is reading these words on this page? Who or what does the thinking and deciding to move on to the next paragraph?

Over the years the answer that seems to best fit the questions above is the mind. We have come to believe that we think with our minds, see with our minds, hold language in our minds, use our minds to make decisions, store memories in minds. We imagine things in our mind's eye where mind is regarded as a faculty of the person. We are less sure that other living organisms have minds (especially as they fail to articulate what is in them).

Once we have separated out the sense organs from the senser the problem has been how to get them back together again. How do body and mind interact and interface?

Descartes was at least bold with his hypothesis that the 'mind', which he assumed controlled the 'body', resided in the pineal gland. Such body-mind and all other dichotomous mind-matter dissections of living organisms have proved to be a big distraction in philosophy and science.

The perennial (false and never to be resolved) problem posed as the question of the relation between mind and matter is one of the unicorns of philosophy and psychology and nowadays of neuropsychology. Elliott Jaques

Once you have split the whole organism into two parts, one called the body and the other called the mind you can look everywhere but nowhere will you find a separate part called a mind.

A body is not a part of a living organism; it is the whole of a dead carcass. Elliott Jaques

The dichotomy has led the search for the seat of living activity to the inside of the body, like Descartes and the pineal gland, and continues to this day albeit in more sophisticated ways.

Gilbert Ryle has clearly argued that the mind-body or mind-brain dichotomy is to begin with a serious logical category mistake, one that makes attempts to reduce what he calls mental conduct to mechanical explanations simply absurd.

Does it matter? The mind-body problem has been floating around for 2500 years unresolved. Are there any real consequences that make it worthwhile to address this thorny philosophical subject? Actually it matters a great deal.

A serious category mistake has created problems in medicine where we treat the symptom and not the patient. In work, we slice and dice people using spurious dimensions and lose the essence of the person in the process. We soon lose sight of the amazing capacity of the total organism to make sense of its surrounding and act without full knowledge. We can make much more sense of the behaviour of the whole person when we relate this to the environment in which they live and the job that they do than the mind being in control of the body.

What is the solution? In the terms of Personal Construct Theory that I introduced in Chapter 1, 'mind-body' (and all its variations) is a polarity that has run its course. It is a polarity that we have invented to help us make sense of what we do. It is not reality. The solution is simple and effective. Ignore it! Without this dichotomy, life and the behaviour of all living organisms makes more sense and there are immediate benefits in all practical areas including how we treat the ill and employ the fit.

We think we know what will make us happy

When we think about the future, we make emotional forecasts. In our life choices we are placing bets on what will make us more or less happy in the future. However, what if our capacity to forecast our future happiness is faulty?

We are not great forecasters about our future selves. We consistently get it wrong. Bizarre as it sounds, we don't really know what will make our future selves happy. When Doris Day sang 'que sera sera; the future is not ours to see' she was only half right. We don't know the future but we still want some insight as to what lies ahead.

Imagination helps us fill in the gaps in our knowledge and is a powerful force that helps in picturing the future. But, just as our memory lets us down about the past, our imagination incorporates some design flaws that make it difficult to get it right about the future. We imagine what life will be like in the future and so often it turns out differently. Even if it turns out as expected somehow it doesn't feel as good as we were expecting.

Chapter 12
Taxonomy

The new understanding of human capability enables the development of a Work Levels Taxonomy that is like a double helix. On one side of the helix is the work requirement and on the other are the individual capabilities that enable the challenges to be met.

This taxonomy is based on Boole's schema first referred to in Chapter 3. Within this there are four types of information processing. Building on Jaques' insight and research, these four types recur at different *orders* of complexity.

Using terminology borrowed from Jaques, these four processing types are named, 'declarative', 'cumulative', 'serial' and 'parallel'. As you go from one to the next there is a manifest increase in complexity. Once you reach parallel processing the next step involves a return to a declarative process in the next order but now the information itself being used changes. It moves to a new and higher quality in its complexity. Information is at a different order of abstraction.

The difference between humans and the information processing of the primates is that we are operating at a higher level of abstraction and this is communicable through language. Language provides the facility to articulate our knowledge of the world in more complex and general terms. This facility allows adults to work at a level of abstraction that naturally includes types of activities, classes or categories. Examples include information about investment and investment includes intangibles like money or rent. Some adults can also work at a higher level of abstraction in which abstract concepts are explained by reference to other abstract concepts.

Each of these information-processing levels provides a specific 'cognitive frame' that naturally encompasses a particular time-horizon. The cognitive frame sets the boundaries for the mental models that we build. At each successive level of information processing, the potential model that a person builds increases in breadth and depth.

People have different levels of cognitive ease within these different types of processing. When they struggle, there is 'ego-depletion'. Ego-depletion opens the possibility of basing actions on lower level cognition and being prone to systematic errors. When there is cognitive ease, the conditions for feeling 'in-flow' are laid.

The Work Levels Taxonomy defines various levels of complexity and makes the dynamic link between these and the sort of work that is required at each level and the necessary capability of the person to handle it.

In this taxonomy, any job in a commercial setting can be allocated to a specific level with its own theme, purpose and core contribution. Core contribution describes the outputs of the job and the value of these to the organisation is in direct proportion to the complexity of the environment in which decisions have to be made. These levels cover all the work undertaken in an enterprise from the highly routine and repetitive to the most varied and strategic. This work spectrum corresponds to a shift in required perspective as the key question moves from 'what do I need to do now' to 'what will we be like in years to come?'

In creating a dynamic link between jobs and people, the taxonomy implicitly recognises that the job accountabilities set the context for the person to operate within. This context will shape a person's approach. It also recognises that the person shapes the job.

The Work Levels Taxonomy provides the basis for optimal organization design in which people match the work that is required.

Level 1) Information Processing Type; Declarative

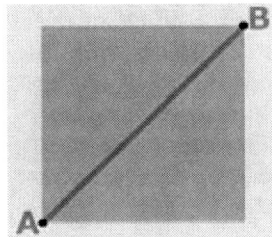

Example; I would do X because of A.

In situations where there is a concretely specified output and method layout, a person using this capability would argue as follows; if trouble A1 do A2; if trouble B1 do B2. If no success; see manager.

Ideal Use:
- Applying skilled knowledge,
- Selecting a known option.

Time Horizon; Up to 3 months.

A person with this capability is likely to make their best contribution when the main focus is on the task in hand. This requires the mental capability to attend to the specifics of a situation in the here-and-now and place this within the context of a workflow.

This relates to a specific theme of work; Task Level.

The purpose of a job at this Work Level is to produce product and/or deliver service within specified limits. The work directly relates to the reduction of costs by prioritising tasks and correctly utilising tools, methods and processes. It is the foundation for customer satisfaction. The people operating at this Work Level are the eyes and ears of the organization, either in terms of quality of output or customer satisfaction.

The danger is that people who lack the necessary 'here and now' awareness miss quality targets. This can be due to having a poor attention span or being over-qualified for the job. Also, over-specifying tasks may reduce the scope for a person to provide vital quality and customer service information. Alternatively, supervisors and line-management could ignore vital feedback.

Decisions are often based on instant interpretations with the dangers of bias.

Level 2) Information Processing Type; Cumulative

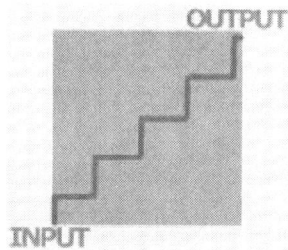

Example; I would do X because of A, B and C.

In situations where it is necessary to diagnose root causes, a person using this capability would argue as follows; if note A, then later B, then later C; conclude problem X and take action to prevent reoccurrence.

Ideal Use:

- Diagnosing a problem within a known type,
- Deducing – (if this and this is the case then this follows),
- Extrapolating (given this and this, then this follows).

Time Horizon; 12 months.

A person with this capability is likely to make their best contribution when the main focus is on case-by-case issues within processes. This requires a depth of thinking and the mental flexibility to link factors, diagnose issues and to draw conclusions.

This relates to a specific theme of work; <u>Process Level.</u>

The work directly relates to managing people and/or working on a case by case/project basis to reduce costs, ensure safe operation, maximise performance and exploit opportunities.

The major added value at this Work Level is the quality of in-depth analysis, problem solving and customer focus.

The danger is that people will pursue perfection or the complete answer when the results don't warrant this. Also, people who are good at trouble-shooting often like the buzz associated with this. They may lose interest in fixing the real issue and take expedient short cuts that cause other problems. Specialists may believe that a technical answer speaks for itself and fail to connect with the customer or user.

Decision makers can sound rational but this can hide a range of bias.

Level 3) Information Processing Type; Serial

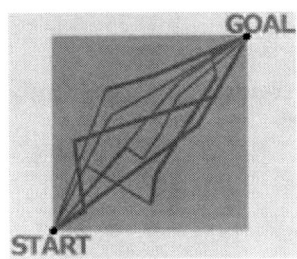

Example; I would do X because A leads to B and this leads to C.

In situations where it is necessary to take a serial approach, a person using this capability would work out alternative plans for getting from A to goal B via L or M or via P and Q and choose the best one.

Ideal Use:

- Planning,
- Contingency making.

Time Horizon; 2 years.

A person with this capability is likely to make their best contribution when there is a need to focus on a goal and achieve balanced results by connecting independent factors into a coherent system and identifying contingencies.

This relates to a specific theme of work; System Level.

The work directly relates to finding opportunities to do more of what is done well and to increase effectiveness, minimise waste and maximise cost/benefit ratios. It inputs to business plans/strategic debates and acts as the lynchpin between the strategy and operations.

The major added value at this Work Level is the establishment of efficient and effective systems, units and functions.

The danger is that a predominant emphasis is placed on standardisation and control. Whilst these are very important elements, when they become the heart of a strategy, opportunities to innovate can be lost.

A person at this level is looking for patterns and preferences but could misinterpret these.

Level 4) Information Processing Type; Parallel

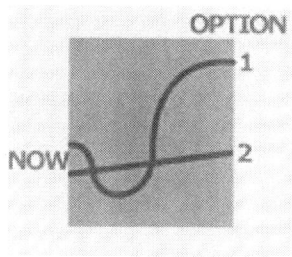

Example; I would do X because of A but you need to consider the impact of X or the impact of not doing X.

In situations where it is necessary use a parallel approach, a person using this capability would use critical path analysis to coordinate a number of serial processes to arrive at a control point.

Ideal Use:

- Modelling.

Time Horizon; 5 years.

A person with this capability is likely to make their best contribution when the main focus is on a major function or unit within its market context. This requires a breadth of thinking and the mental capability to model options that can create the possibility for significant change.

This relates to a specific theme of work; <u>Breakthrough Level.</u>

The work directly relates to building long-term competitive advantage and removing value-destruction. It has to successfully translate strategic intent and purpose into clear goals.

The added value of this Work Level is to create strategic options and bring about change.

The danger is that people get caught up with the enthusiasm of change and innovation and under-estimate the time that it takes to align effort, translate ideas into actions and manage company politics. Alternatively, they can go off on a personal tangent; pursuing ideas of individual interest. Another possibility is that someone uncomfortable with strategic work may hide their unease by focusing largely on today's customers and miss out on emerging markets or future customers.

The above types of processing detailed in Levels 1 to 4, repeat but this time at a higher level of abstraction.

Level 5) Information Processing Type; Abstract Declarative

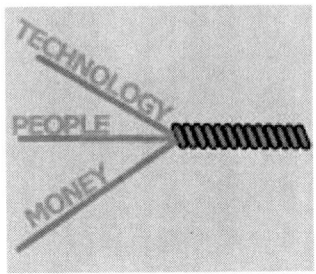

Example; I would do X because of abstract concept A, or B or C.
Ideal Use:
- Setting the context within broad socio-technical and political context.

Time Horizon; Up to 10 years.
A person with this capability is likely to make their best contribution when the main focus is on the complete operation at a national level or within an international context. This requires a breadth of thinking and the mental capability to imagine a compelling and different future.
This relates to a specific level of work; <u>Strategic Intent.</u>
The work directly relates to building value-chains and increasing the net-worth of the intangible assets.
The added value of this Work Level is brought about by the quality of the communications - both internally and externally. Through this, the strategic path is set and the means for delivery are established.
The danger is that skilled communicators begin to believe their own PR or they have too optimistic a view of human capacity to change and do the 'right thing' without being set clear accountabilities. When a jobholder at this level fails to understand the interconnections that are required, the result is that resources can be over-stretched and risk-analysis ignored. The flipside is having someone operate at this level who has the ability to develop the conceptual framework but fails to communicate the need for change or make the link from innovation to the business imperative.

Level 6) Information Processing Type; Abstract Cumulative

Example; I would do X because of abstract concept A, B and C.
Ideal Use:
- Reframing.

Time Horizon; Up to 20 years.
A person with this capability is likely to make their best contribution when the main focus is on a set of operations often within an international context. This requires a breadth of thinking and the mental capability to bridge highly diverse and conceptual aspects in a way that crystallises the issues without overly simplifying them and revealing new perspectives.
This relates to a specific level of work; Citizenship.
The work directly relates to building and sustaining the goodwill of all stakeholders and ensuring that the group has a 'licence to operate'.
The added value of this Work Level is setting a path and context for the leaders of businesses, strategic business units or research and development establishments operating across diverse geographies and markets.
The danger is that without the right mandate and with the wrong incentives, leaders at this level can increase, over the short-term, the share price of an enterprise whilst destroying long-term value.

Level 7) Information Processing Type; Abstract Serial

Example; I would do X because abstract concept A leads to abstract concept B and this leads to C.

Ideal Use:

- Holistic thinking and creating new possibilities.

Time Horizon; Up to 50 years.

A person with this capability is likely to make their best contribution when the main focus is at an industries level within a global context. This requires a global breadth of thinking and mental capability to preview consequences on a global scale.

This relates to a specific level of work; Global Transformation.

The work directly sets the value system in operation.

The added value of this Work Level is the picking up on small signs of future possibilities and nurturing these into being the industries and social norms of the future.

The danger is that the shareholders, stakeholders and power-brokers don't give time to leaders unless they are seen to be miracle-workers.

Level 8) Information Processing Type; Abstract Parallel

This is the level of the true genius and, to all intents and purposes, can only be measured by others of similar capability.

It is entirely possible that there is a whole order of complexity that the rarest of humans can work at above these identified levels.

Part Three
Standing on the Shoulders of Giants

Chapter 13
The Irreducible Mystery of Life

Life is an irreducible mystery but not if you follow many evolutionary biologists down a reductionist path.

If you flick through this book trying to get a feel for it you are likely to sense that it is about the whole person in a work context. What, you might ask, is he writing about evolution for? Is the author throwing in the kitchen sink to pad out what otherwise might be a thin book? No. The theory of evolution is a huge idea that shapes our sense of who we are and what we add up to and it also serves as an important analogy in our business age.

In shaping a sense of who we are, current theories of evolution have missed out the vital ingredients of freedom and choice in all living organisms. The consequences are very real and detrimental. The theory lends itself to those who promote that life is essentially meaningless and futile. It also gives weight to those who believe that, as competition is the driving force of evolution, an aggressive 'winner takes all' must be OK for society too.

There is nothing to argue about with Darwin's theory that evolution happens by means of natural selection. Evolution happens. Sometimes it happens fast as in the case of dogs and flowers when humans select for particular traits and characteristics and sometimes it happens slowly over deep time as organisms adapt to the environment. However, once you move on and identify the motor of change as the selection of advantageous genetic mutations that replicate themselves through the gene pool *and* you identify the information storehouses and the learning mechanisms (*and* make the mistake of identifying the brain as the place which decisions are made), the end of this process is to see mankind reduced to a series of mechanisms and processes. In the final analysis we become simply the means by which DNA replicates itself. This is the territory of the 'selfish-gene' as first expressed and elucidated by Richard Dawkins.

When you take on board all of that amounts to mainstream evolutionary theory and follow Dawkins down his particular reductionist path towards the idea of the selfish-gene you inevitably come to see life as essentially futile.

Dawkins' selfish-gene argument goes like this; natural selection is all about the survival of self-replicating instructions for self-replication. In his book, *The Greatest Show on Earth*, Dawkins illustrates his idea by referencing anacondas. Anacondas are built by coded instructions whose ultimate message is, like a computer virus, 'duplicate me'. In the case of a computer virus the instructions are carried out instantaneously. In the case of the anaconda there is a huge digression as part of the efficient execution of this fundamental message. That digression is the anaconda and the life of the anaconda. The DNA message is 'duplicate me by the round about route of creating an anaconda first'.

If a variant of DNA survives through an anaconda being able to eat Richard Dawkins whole then that is all that is needed by way of explanation.

Can we really believe this reductionist message? No. No. No.

Evolutionary biologists are now adept at fielding the criticism that they are reductionists. They say that evolutionary theories of behaviour, including altruistic behaviour, are not necessarily *caused* by genetics. So long as the behaviours in question have a genetic *component*, i.e. are influenced to some extent by one or more genetic factors, then their theories can apply.

However, in loosening their language from *cause* to *influence* most are paying lip service to the interaction of the environment and heredity (as manifested in the organism). Clearly, there are important roles both for heredity and the environment but making the interaction of the polar opposites to be the cause of behaviour leads to the organism vanishing; can there be an environment without someone to perceive it, a behaviour without a 'behaver', learning without a learner? As Alice thought to herself when the Cheshire Cat vanished quite slowly, ending with a grin which remained some time after the rest had gone, 'Well I have often seen a cat without a grin but a grin without a cat! It's the most curious thing that I have seen in my life!'

Dawkins and fellow thinkers have been focusing on important mechanisms but have failed to see the impact of another system in operation. They have failed to take into account the agent in the process and the capacity of an organism to make choices; choices that are bounded by the capacities bestowed by DNA and physiology but are not determined by them.

Seeing the organism as a clever way of the gene replicating itself belittles the organism in question. Seeing life through the eye of the gene makes for a good story but constitutes a depressing philosophy. In providing a non-reductionist perspective, it is vital to make the distinction between three different sets of functions, namely the phylogenetic, the physiological and the ontological. Phylogeny comes from the Greek word for the birth the species (phylum) and ontological comes from the Greek word, ontogeny, for birth of being.

The physiological functions include the neuronal, immune, endocrine, digestive, circulatory, muscular and reproductive subsystems. These subsystems work together to ensure that an organism remains alive and each influences the behaviour of the organism as a whole. However, there is not a one to one relationship with the working choices made by the organism. Even the neural subsystem including the brain and particular parts of the brain, whilst necessary for certain behaviours, are nevertheless not sufficient to explain particular judgements and choice behaviours. They provide a field, which may be broad or narrow, within which the choices are made.

The phylogenetic functions also contribute to the field or context within which organisms use judgement and make choices and decisions. These are the functions that are genetically found in the species. They persist over thousands of years and change under evolutionary pressure to adapt to the environment. Under this category we can include a vast array of actual behaviours that come under the heading of innate or instinctual such as the nesting of birds, the dam building of beavers, the animal migrations, hunting, herding, acquisition of language in humans and all other more or less well known programmed behaviours.

These behaviours are related to particular gene systems and, as such, are understandable in terms of molecular biology. Just as with the physiological subsystems, the gene-driven behaviours set the direction of the behaviour, they do not and cannot explain the detailed judgements, choices and decisions under the particular circumstances prevailing at the moment. These choices include where to put a nest, which available material to gather to build a dam, deciding on whether the river is running too fast to cross it in the migration or in the case of humans which words to articulate. Coming back to the Anaconda referenced above, the phylogenetic and physiological functions explain everything until the point is reached for the Anaconda to choose to eat Richard Dawkins or not.

These choice-making behaviours comprise the third function, which we can call the ontological functions. The ontological functions bring the organism centre-stage and in so doing we will no longer see behaviour as simply an output of the interaction between genetics and the environment. The *being* is an active agent in its own right. The ontological functions are the ineffable processes that I referred to in Chapter 11. They constitute the fine-tuning of decision-making work within limits including innate instinctual demands and, in the case of humans, limits set by linguistically constructed rules, regulations and values. The fine-tuning is always there, always essential.

The fine-tuning is the proactive part of decision-making. It requires the setting of goals, either intuitively or in verbally expressed form. For that to happen there has to be a construction of the future, that is, the setting of something to be obtained or somewhere to get to at some time. This time-horizon could be milliseconds or in the case of humans, days, months or years.

These fine-tuned choices and decisions are not programmed. They are idiosyncratic. They are not fully linked or linkable to any parts of the organism; not even to the brain in organisms that happen to have a brain. They are the products of the dynamic functioning of the whole organism and that, may I say, is that. The results of the choice process are readily observable in the actions that follow but how the choices were made is not accessible. This ineffability, this hidden nature of the choice process is what we call free will.

To really show how the free will component is core to life itself I'll focus on the amoeba and its goal directed work. Through this unicellular organism, I hope to demonstrate the nature of its work in the world uncluttered by having to take into account the function of the brain or the use of language.

Before we begin, I would like you to think carefully of the linguistic interpretation that you make of words like 'choice' and 'choosing'. I am going to consider what the amoeba must know in the sense of *knowledge in action* (rather than in words) that enable the amoeba to do all the remarkable things that it does. Just because I describe them using linguistic propositions, and the amoeba cannot, does not mean that the amoeba hasn't the necessary primal awareness and knowledge. As such, I would like you to place your linguistic overcoat on the peg and surround yourself only with what is necessary to understand the private life of the amoeba. In this I am referring to the research undertaken by the esteemed biologist Lynn Margulis (1938-2011).

What can be observed is an amoeba purposefully engaged in seeking things. It is not random in its behaviour or its likes and dislikes. It finds *Tetrahymena* delectable but is not so keen on *Copromonas*. It squirms forward on retractable pseudopods and chooses between tackling a *Copromonas* just at the end of an extended pseudopod or to take on a more difficult vector in pursuit of a tasty *Tetrahymena*. It deals with entities as it moves through its world, taking some and pushing others aside. It chooses. To decide between this and that it must have, however primitive, a model of its world and attend to this. To make the sort of choices that the amoeba makes it must have a usable awareness of space and direction otherwise it wouldn't be able to locate itself and potential foodstuffs around it. The amoeba is making important distinctions in terms of properties including weight and size. It doesn't expend energy on choosing to try to eat something that is too big for instance.

You may be nodding your head up to this point but here is a tough one to get hold of; the amoeba must have an awareness of probability. It is not a probability that it can articulate to us or to itself but it is a calculation that leads it to choose between the entity that is easily caught and one that is probably not so desirable anyway. What is more, the amoeba must have an innate awareness of cause and effect. This is an anticipation in the form of (but so embedded in the

90

organism that it doesn't need to articulate it) 'if I want to go there, then if I do this (cause), I'll be well on my way (effect). Linking cause and effect in this anticipatory manner is the absolute foundation stone of the reality testing necessary for survival in all living things.

Interestingly, certain types of amoeba (Mycetozoa) are social creatures. In certain conditions they are surrounded by countless members of the same club all of whom communicate with each other using self-explanatory chemical signals. Using the chemical combinations from one to another they form a 'slime' or common colony that enables them to collaborate and move, en masse, to damper and richer feeding grounds. Any such signalling system must presuppose a primitive mentalese; what we can call, in this case, a universal amoeba mentalese.

This glimpse into the private life of the amoeba shows that it is just like all organical systems, i.e. continuously engaged in making judgements, choices and decisions that require a sense of greater or lesser probability in any given case. The life of certain amoeba includes a social world in which all are locomoting; all moving to a purpose; all making choices in the absence of knowing exactly what that choice will bring.

The amoeba is a unicellular organism and not that primitive in the great scale of things. What about bacteria? Well, it is becoming increasingly apparent that bacteria are capable of communicating and cooperating in the same way as the amoeba. What is clear is that all mobile microbes when offered food make selections; in other words they choose. It is not far fetched to say that they have some sort of intention. They cannot form these intentions and articulate them but they are getting on with them anyway.

What is clear from close study of primitive organisms is that they are engaged continuously in the work of choice-making, problem-solving and cause and effect reality-testing in order to achieve a goal. These aspects cannot be reduced to genetically driven behaviour even in the most primitive organisms.

Did you leave your linguistic overcoat at the door and utilise only what is necessary to understand the private life of the amoeba? If you think that I am attributing human-like capabilities to things that are simply encoded to react to the world in a particular way, you surely brought in a range of assumptions that are simply not necessary to fully understand that all living organisms have the capacity for direct engaged awareness and through this; free will.

If you think that I have left the planet, please move on. There are some very down to earth implications of the new model of capability for students of economics and practitioners of human resource management.

If you have followed me so far, then I encourage you to take one step further and take a close look at the margins of life where DNA holds sway. Here DNA sends out its messengers as RNA to communicate, yes *communicate,* with the ribosomes mobilizing them to produce the amino acids and make the protein out of which to build people like you and me. Is there room here for choice and intention? A chemist might conclude no; these are simple 'click and go' chemical processes and need no further explanation. Yet, this is far from simple chemistry. The whole quality of the organizing and functioning of this chemical matter is at a different level of complexity from simple chemistry.

I am not suggesting that a spiritual force, magical power or animal spirit needs to be added to move the chemistry on. As we have already talked about in previous chapters, information is embodied and here at the margins of life it is embodied in DNA. The suggestion made by Jaques in his book, *The Life and Behaviour of Living Organisms* are that, at the level of DNA, matter is *already dynamically organized to function in living form* and, at a 'proto-level', choice and intention are built into the information processing system. As such, choice is enmeshed in the fabric of life and does not spring into being at a particular level.

To progress further, these ideas could usefully be expressed in the language of mathematics and therefore I will have to leave it here.

Thank you for venturing with me into the margins of life. Back now to firm ground. DNA and physiology provide the field or context or boundaries within which free choice operates. The absolute final step in the work process is the choice between this *or* that and in making the choice the organism will find itself on its way to its goal. That is true of amoebae and it is true of you and me too.

As such, behaviour is not wholly determined. There is a gap that needs to be bridged by intention, a moving forward to achieve a purpose.

It is this gap, characterised by the absence of complete information in the world, that current evolutionary biologists especially those persuaded by the writings of Richard Dawkins fail to understand. The gap can only be filled through an organism having intention and making choices. Even with the strongest genes, the healthiest of immune systems, the most powerful learning mechanism and readily available received wisdom, the organism could not step into the gap. The programme or mechanisms identified by Dawkins et al. would fail. Life is not unrelated to physics and chemistry but it cannot be fully explained by them. The life process that includes the '*beingness*' of an organism cannot be reduced to physiology or via physiology to chemistry or physics. These life functions, the needs, the intentions and goals, the observations and attention, language and communication, the choices and judgements are not related to any specific physiological parts, genes, tissues, fluids or cells.

Rather than seeing the organism as a clever way of the gene ensuring its survival, it is much more life affirming when we understand that the organism, through the capacity of choice, is the agent in the matter of its survival.

The essence of *being* is a whole organism process and is the mystery of life. It will always remain so. When we fully appreciate that living organisms are all irreducible mysteries we might not be so ready to see life as futile. Not only that, we can begin to question the received wisdom that comes with the theory of evolution and it is that to which I turn in the next chapter.

Chapter 14
Please Send Answer Via Twitter

Have I mentioned that I have an 'A' Level in biology? I am justifiably proud of this minor achievement but recognise that this does not put me in a position of being an expert when it comes to evolutionary biology. A little bit of knowledge in the hands of the wrong person can be dangerous so all I can do is speak from the position of the educated man on the street.

What this particular man on the street is faced with is a choice between believing that God designed all living organisms or natural selection enabled the evolution of lower organisms into higher ones or that there was a big bang and out of chaos and randomness suddenly there was wondrous diversity.

When you understand that there is only one rational option, evolution by natural selection, the man on the street also absorbs a lot of other received wisdom. Summing it up in less than 140 characters suitable for the medium of our age; evolution happens because nature 'is red in tooth and claw' and is the place where only the fittest survive.

The last hundred years or so have seen the systematic application of Darwin's theory to all aspects of life without exception. Core to evolution are the ideas of competition and natural selection through out-producing others. The tough, no-nonsense, highly competitive role models that we take as the captains of industry give testament to this prevailing view.

If you are an evolutionary biologist don't worry, I am not going to say 'show me the evidence, where is the missing link?' There is plenty of evidence about missing links. There will be no argument from me that evolution is not for real. However, I am making the very important critique that mainstream evolutionary theory does not fully explain the development of cognitive complexity.

I'm not suggesting that we bring back the notion of divine creation nor will I introduce ideas of intelligent design. Evolution through natural selection can explain the existence of a complex organ like the human eye. If not God or natural selection what else could be at work here? Not randomness of a big-bang type. Creating the human eye through big-bang randomness is as probable as a hurricane blowing through a scrap-yard and assembling a Boeing 747. And there is no need to say that *if* given enough hurricanes etc. No, it is not going to happen. Evolutionary biologists convincingly argue that the human eye sees well now because it came from a long line of descendents that saw a bit better than their rivals which enabled them to out-produce them.

Mainstream evolutionary biologists describe the underpinning mechanism for adaption to the environment (which includes the development of the human eye) in the following way: change is built in; genetic mutations happen at a constant rate. These changes are subject to a ruthless cost benefit analysis; change that is too big is discarded, change that is too small is ignored. In natural selection there is no intention or foresight; redundant capability is not built-in because it might be useful in the future. In summary, small random changes in the underpinning genetic code, sufficient to enable the holder to add more of their beneficial genes into the gene pool than their neighbour, are the incremental steps of natural selection.

Here endeth the lesson. These are the rules in mainstream evolutionary theory. I'm not going to argue with the rules. I only want to show that the rules don't seem to work all of the time.

In order to fully understand where the rules apply and where they don't, we need to move away from the image of evolution being like a branching tree. The single, branching tree model suggests that the whole process is moving upward, which reinforces the notion that the higher branches are somehow the better branches and, of course, the Homo sapiens branch is the best of all.

There are two evolutionary directions; there is a lateral direction and a vertical direction. Neither of these directions is better, only different. Vertical movements signify changes in degrees of cognitive complexity and lateral movements signify length of time of survival on earth. If you have to choose which direction is better you could point to Homo sapiens at the top because of how clever they are. Or you could choose the lateral direction and point to the bacteria for instance that have survived three or four billion years and have now successfully colonized every other living organism including ourselves. Each of us provides board and lodging to some one hundred trillion bacterial residents.

If the image of the tree has any value then we need to see this as involving several trees each of them in the unusual posture of lying down and largely stretching sideways in search of survival rather than reaching for the sun. The trees lying sideways represent the successful survivors who have managed to bend with the winds of change and have continued to evolve over millennia within their level of complexity.

The mechanism for the progressive adaption to the environment in the lateral direction is clear. At the heart of this mechanism is genetics and in particular DNA. However, the mechanism of adaption in the vertical plane (cognitive complexity) is not so evident.

This vertical direction is best understood when we use the model of information processing that was introduced in Chapter 3 and developed further in the Chapter 11. In this there are four basic types of information processing that repeat at progressively higher orders of information complexity. This provides a measurement scale that has been lacking until recently for assessing where organisms are in the evolutionary scale when it comes to cognitive functioning. In effect, the scale measures the time-frame of intention and this varies from microseconds in primitive organisms to days, weeks and even years in the case of human beings.

Using this scale we can hypothesise that a move from one category of complexity of processing to the next more complex level is an evolutionary stage. Jaques in his book, *The Life and Behaviour of Living Organisms* suggests that nine vertical evolutionary stages have occurred in the long lineage from living matter to Homo sapiens. I won't explain them all here as they are clearly described in his book. The important thing to understand is that at each stage of vertical evolution there is a progressive loosening of the grip of the phylogenetically established instinctual behaviour and a corresponding increase in the freedom of choice in problem solving.

The last stage is particularly interesting as it concerns the appearance of Homo sapiens around 200,000 years ago. This saw the emergence of the ability to disengage, detach and reflect, which complemented the capability for active direct engaged awareness shared with all living organisms. This was accompanied with a move from communication by signalling to the use of language.

This big evolutionary step threw up a species (Homo sapiens) in which, at the level of fully matured adults, there is a capability range that spans four information-processing categories within, at least, two orders of information complexity. This mighty range of some eight or more discrete levels enables a small percentage of the total population to have a time-span of intention (thinking forward) that encompasses decades.

This massive evolutionary change happened over a short period. If we envisage the whole of evolutionary history to be twenty-four hours, the emergence of the Homo sapiens with their cognitive capability fully formed would be the equivalent of much less than one tenth of a second. That is one 'big bang'. But big bang change doesn't easily fit the rules that I detailed earlier in the chapter.

Evolution does happen remarkably quickly on occasion but at the time of the biggest changes underpinning the greatest evolutionary leap in the history of evolution, genetics massively loosened its grip on behaviour.

You will recall from the rules that there is no intention or foresight; redundant capability is not built in because it might be useful in the future. Yet, in the case of Homo sapiens, plenty of information processing power was built-in reserve, which would not serve the majority and the reproductive benefit for the individual 'user' may well be negative. History shows that society tends to do nasty things to those that are seen to be too clever. Ask Socrates, who was found guilty of both corrupting the minds of the youth of Athens and of impiety (not believing in the gods of the state) and subsequently sentenced to death by drinking a mixture containing poison hemlock; ask the intelligentsia of Stalin's Russia or Pol Pot's Cambodia whether it paid off being smart.

Just as intriguing is the fact that in the case of some high capability people, maturation of thinking capability continues well after individuals are at their reproductive peak! Here is potential capability that doesn't lead to an individual out-producing its neighbour and thereby strengthening the gene pool with its specific genetic variation.

In summary, *the* gene-based mechanism for natural selection doesn't stack up. It works for lateral (physical) evolution but not entirely in the vertical (cognitive) plane. **Something else is going on**.

Why, you may ask, am I rattling the cages of evolutionary biologists? They are probably not paying attention anyway and have discounted my views as soon as there was no PhD tag on the end of my name.

I am not saying that the King is partly dressed just for effect. It really matters. It matters for two reasons, one relates to the pursuit of the best theory to fit the facts and the other relates to what we take from science that becomes the secular beliefs of our age. If something at the heart of evolutionary theory is missing why should we be so ready to take on the motto of *survival of the fittest* and use this as the analogy for our age, which does not easily accommodate ideas about collaboration, cooperation and especially altruism.

The story of evolution, as told, lends weight to the idea that it is right and proper that an individual, corporation or nation can take anything that they want. Might is right in the world in which only the fittest survive.

It is much more sensible to see that life preserves life but if and only if the free will judgements lead to good choices. Good here can simply be taken to mean the sort of individual survival that contributes in turn to the survival of the species.

If evolutionary biologists have read this far and feel that I have either misunderstood the theory or stated the blindingly obvious please could they summarise the current evolutionary ideas in one hundred and forty characters or less and send via Twitter so that the educated man on the street can understand it. Many thanks.

Chapter 15
A Copernican Revolution

Nicolaus Copernicus (1473–1543) was a Renaissance astronomer and the first person to formulate a comprehensive heliocentric cosmology, which displaced the earth from the centre of the universe.

Copernicus' epochal book, *On the Revolutions of the Celestial Spheres*, published just before his death in 1543, is often regarded as the starting point of modern astronomy and the defining epiphany that began the scientific revolution. His heliocentric model demonstrated that the observed motions of celestial objects could be explained by putting the sun at the centre of the planets' orbits. Whilst his work stimulated further scientific investigations, the displacement of the earth from the centre of the universe created great controversy. However, with the passage of time it has become a landmark in the history of science that is often referred to as the Copernican Revolution.

A similar revolution is quietly taking place in the consulting rooms of psychotherapists and counsellors and is ready to spread further afield into coaching and change management.

Psychotherapy/Counselling

What is absolutely clear and universal is that we, that is, all humans have a deep-seated need to avoid the anxiety of uncertainty.

Our huge dislike of uncertainty is evident in the following thought experiment. You take a test to indicate if we have a dangerous genetic defect. The result can be:

1) Conclusive; you don't have it,
2) Inconclusive; you may or may not have it,
3) Conclusive; you have it.

How do we feel about those outcomes? Yes, you feel most anxious when the test is inconclusive. Human nature can withstand lots of things but uncertainty isn't among them. Uncertainty keeps us in a state of anxiety and our mental software is not well equipped to deal with it.

We want to know what happened and why it happened. We just need to know something, anything, that makes sense of our past, present and future.

This is where psychotherapists enter.

Referencing the Copernican Revolution is entirely appropriate because we all place ourselves at the centre of our universe and standard approaches to psychotherapy reinforce this notion.

The main theories of psychotherapy including psychodynamic, behavioural and cognitive ones are attempts to describe human experience within a given framework that provides explanation of the way things are and the causes that have brought the situation about. The end result is that they isolate individuals in a self-oriented universe of their own making.

Psychotherapy as a profession is remarkably united in its emphasis on the individual-centred concepts and concerns. Most current forms of psychotherapy have the common objective of cure through an understanding of human conflict in intra-psychic (e.g. id, ego and superego) and interpersonal terms. The cognitive-behavioural emphasis on social management of one's mental resources is a pragmatic variation on the same theme.

A 'benefit' of psychodynamic, behavioural and cognitive theories is that they enable experts to quickly judge others. They provide the definitions that allow us to categorise people as mad and sane and allow us to diagnose 'pathology' and 'trauma'.

The ubiquity of pathology and trauma in psychotherapeutic systems is not unlike the omnipresence of sin in religious systems. The more there is of these conditions the greater the need for therapists. This is similar to the way that sin demands the need for the clergy. Van Deurzen

Psychotherapeutic culture has helped us turn away from old-fashioned religious notions of God and the battle between heaven and hell. Instead, it has provided for the salvation of humankind through coming to terms with ourselves and the raging forces in our own psyche. This neatly enables the psychotherapist to become a kind of secular priest who is not accountable to any higher body (other than the strict order into which they have been granted communion). In this powerful position they exert huge control. They identify pathology and then seek to remedy this through interpretations or interventions that give a causal explanation for what ails. In the process, the client's perceptions are reframed and rearranged in such a way as to make the therapist the central power in the process of being saved from ourselves.

The 'contract' with the therapist is clear depending on their specific orientation. Psychoanalytic therapies promise a life of ordinary human misery but this is hardly a satisfying replacement for the salvation promised by religion. The cognitive-behavioural model promises effective social functioning but it overlooks the yearning of the individual for more than just being able to obey the social etiquette of our time. The humanistic or positive psychology movement promises self-actualisation and comes close to being inspirational but fails to address the dark and sometimes evil side of humans on earth.

Whether intentional or not, mainstream psychotherapy reinforces a sense that we are the centre of our universe. We have this sense because we are all, egoist and Mother Theresa alike, the focal point of a network of interactions. I might not think that everyone is there to serve me but the centre of my experience is always in myself. However, this 'mineness' is profoundly problematic for it is generated by and dependent upon my connection to that which is not me. It is often hard for people to grasp the lack of substantial reality in ourselves as essentially we are nothing without our relationship to the world of things, people, events and ideas. Although we might feel solid and have positive self-images and strength of character, we are essentially empty vessels that only come to life in the process of resonating with what we encounter.

Although I am the centre of my existence, my being is not a substantial entity. My 'I' is an opening, which lets through the light of existence. I can regulate the extent to which I let things in or keep out but essentially my being is the medium through which life flows and comes to light. Van Deurzen

When we pretend that the 'self' is a substantial entity, separate from the world, we distort our image of what a person is or can be. If we think that a person's inner world can be measured, assessed, sliced and diced as if it is a solid object we make a grave mistake. Human life can only be understood in action, interactively and dynamically.

'The price of rejecting easy answers is that we must be willing to tolerate ambiguity and accept one's own ignorance.' Richard Feynman

Existential psychotherapy provides the basis for the Copernican Revolution. It presents a philosophical alternative to other forms of psychological treatment by emphasising the problems of living and the human dilemmas that are often neglected by practitioners who focus on personal psychopathology. In this form of therapy, human beings are no longer described as at the centre of the universe; we are interrelated with others, with a physical world and with *being* itself.

Existential psychotherapy aims to help people handle ambiguity, conflicting interests and paradox by firstly helping them recognise that these are at the heart of being human. Not immediately knowing the right answer brings with it feelings of insecurity. Whilst 'angst' is an integral part of the human condition, so too is the feeling of relief, even joy, of rising ourselves above issues, gaining insight and moving on. A few gifted people are able to use life's uncertainties as a resource. Through working with dilemmas and paradox they

101

transcend dualist dimensions and create an encompassing new perspective.

Existential psychotherapy is different to all common therapies because it takes as its subject matter uncertainty, anxiety and doubt and does not try to relieve these through identifying causes or treating symptoms. It recognises that these might feel like the curse of living but they are also the source of our freedom.

Coaching and Change Management

The Copernican Revolution has yet to spread to the world of coaching and change management. In this multi-million pound industry there are a plethora of tools and techniques used to analyse the individual as if they are a substantial entity unaffected by the context in which they operate.

The unbearable lightness of being that sometimes afflicts us explains the popularity of personality 'tests' like Myers Briggs. In the case of Myers Briggs, it uses four dimensions to generate sixteen categories that the whole of humanity can be allocated to. These categories neatly explain personal preferences and, on the basis that preferences are likely to be practiced, 'experts', by sleight of hand, go on to predict how people *will* behave. I am continually surprised when people fill in a questionnaire and then profess amazement that a very general description apparently matches their self-image. For me, it demonstrates that our self-image is very fragile. We are constantly in the process of forming and reforming it. When a 'scientific' test proves that we have a certain personality type many of us grab on to the language that makes this real. For some, the description becomes a self-fulfilling prophecy as they start to behave in the way that the profile suggests and explain this as, 'well I am an ESTJ after all!'

Over the last thirty years or so there has been an explosion in psychological profiling techniques that purport to measure everything from intelligence to aptitudes for all types of thinking and physical skills. These largely cater for people's desire to quickly put a label on others and categorise them as this or that kind of person and then to use this knowledge to predict their behaviour. However, this process of labeling, without the effort required to really know the person, doesn't take into account context - but human beings are highly context sensitive. As such, saying that somebody is highly extrovert, for example, is far too broad. Rather it is necessary to understand the situations in which they feel they can be extrovert and the ones in which they feel less so.

To fully appreciate a person, it is necessary to take a rounded view and always look at them in terms of the context of the work that is required.

Instead of focusing on personality and context free traits and factors it

102

is necessary to address a critical area that is often neglected; namely the breadth and depth of a person's thinking. If 'thinking' is focused on when profiling people it is often limited to the classic aptitudes related to logical reasoning and analysis. However, a person's perspective on the world and the thinking that gives rise to this is highly relevant as:

- Our ability to think, especially our ability to think about thinking, defines us as human beings,
- Thoughts and words are incredibly powerful, - they have caused many of the great unions and the great conflicts throughout history,
- Awareness of our thinking intentions is real self-insight,
- We value what we think about and we think about what we value,
- We are not trained to think strategically, yet this is key to 'direction-giving'.

When thinking is addressed we can identify a person's current level of potential capability. This is the highest level that a person can work at the present time, provided that the work is perceived to be of value and they are given the opportunity to acquire the necessary skills, knowledge and expertise.

In an attempt to create certainty, agreement and consensus huge amounts of time, effort and money can be wasted on change management.

In Samuel Beckett's 'Waiting for Godot' (1956) the characters reflect, plan, resolve and procrastinate but never act. The play ends with the following dialogue:

Vladimir: Shall we go?

Estragon: Let's go

[Stage directions; no one moves.]

This example of decision-making sums up what is happening in many team meetings throughout organizations. Someone makes a suggestion, everyone agrees and an apparent decision is made but, in effect, no one moves.

Games such as these are rife in organizations and manifest in the following symptoms:

- Slow response to market opportunities,
- Silo mentality,
- Poor customer relationships,
- Slow decision-making,
- Too many meetings,
- Negativity.

103

Many approaches to tackling the above symptoms rest on psychological theories that see aspects of business life such as interpersonal stresses, aggression and damaging game-playing as the results of psychological processes involving the unconscious. As a natural follow-on they espouse that in order to change behaviour it is necessary to change the psychological make-up of individuals. The change-programmes that result focus on the individual and ask the individual to change. However, the major problem with this, as I have personally discovered, is that personal change at the level of attitudes and values is incredibly difficult and slow even through prolonged counselling/psychotherapy with experts over months and even years.

In a work context, change can be even more difficult. When the people around you and the systems stay the same, any personal commitment to change soon peters out.

With psychotherapeutic approaches in ascendancy there is a critical need to counter-balance the initiatives that set out to change the person with an in-depth appreciation of the fact that individuals are not islands; they are contingent and dependent. Context is King. Individual behaviour is a result of expectation and the system within which we work largely sets the expectations.

In any change process it is necessary to recognise that people are 'meaning-making' and 'meaning-seeking' beings. People make meaning in all areas of their life; family, social and spiritual. Clearly they also make meaning for themselves and others in a work context. To fully understand people in work, it is necessary to understand this context

The Work Levels Taxonomy describes a critically important contextual factor, namely the complexity of the decision-making environment. Through describing complexity at increasing levels it is possible to create a dynamic link between the challenge of work and the individual capability required to undertake this. This link provides a systemic perspective on being in-flow and managing transitions.

Flow

To use the phrase introduced in Chapter 7, people are likely to feel *in flow* when the challenge of the job is matched by their ability to get their head around it.

When in-flow, people are comfortable with using the capability that still differentiates humans from the mightiest computer, namely an ability to pick up on small contextual cues to use judgement even though there may be large gaps in the data. Making judgements means that we use our mental-model of an expected universe so that we can make a prediction. For everyone, the sophistication of the mental-model has to be at least in-line with the degree of complexity that is faced in the work undertaken. If this is not the case, poor decisions result and people become uncomfortable as they recognise that, somehow, the world does not operate to the rules of the mental model.

We thrive when the challenges of the world in which we work, learn and live draw fully on our capability. People all over the world describe feeling confident, competent, energised, even exhilarated when coping successfully with new challenges. They're going with the grain and things just seem to happen right of their own accord. Their intuition tends to be correct and if there are choices to make, they make them almost without being aware.

Where someone is in-flow, it is still beneficial to think about how development can keep him or her in the zone. Coaches can help people who are in-flow review, articulate and appreciate their current role and subordinate roles as well as helping them make comparisons inside and outside of the organization. It is also valuable for them to fully appreciate the demands at the next level and context in which their manager has to operate.

Without the relevant thinking capabilities a person will feel uncomfortable and out of their depth. This sense of being out-of-flow has important consequences for the individual and the organization.

Being in or out of flow has important business impacts. For example, leaders are making choices that have consequences that can take years to play out. Due to the complexity of the issues being handled, it isn't easy to initially see the quality of the decision making especially as some leaders become adept at hiding their discomfort and unease.

When 'what there is to do' outstrips what we feel capable of doing, we are in a situation of being over-stretched. Our work calls on capability we do not have and we find it difficult to choose between alternative courses, distinguish between what should be done now and what can be deferred. We lose the ability to judge the appropriate moment to act. This leaves us little option but to do the tasks we are familiar and comfortable with and neglect those tasks that we would rather not tackle.

Potential risks of over-stretch can include worry, stress and anxiety through feelings of being unable to cope with the speed/complexity of the decisions to be made. A person may be unable to handle all the relevant factors at once and may consider some issues at one point, some at another but never all together in an integrated manner. Decisions are taken using heuristics (rules of thumb that reduce complex problems to simpler ones that can be solved). Usually, long-term consequences are not taken into account.

When faced with uncertainty that is greater than they can tolerate, leaders can 'displace' their unease and make snap judgements. An attitude of 'just-do it' can be a symptom of hiding unease. Alternatively leaders can handle the same state by waiting for the final piece of data that makes a decision a sure-fire success only to find that the opportunity has been missed.

Over hasty decisions reduce a person's tension because they have at least done something but the costs associated with rework soon mount. Poor decisions represent not only poor investment but also lost opportunity and overly laboured decisions have financial and market related consequences too.

In contrast, when 'what there is to do' fails to challenge what we feel capable of doing, we are in a situation of under-utilisation. In effect there is spare capability that is being wasted. When we are not fully utilised, frustration and boredom escalate. Tasks are increasingly perceived as chores as our attitude becomes more and more negative and is characterised by such statements as, 'why bother to think about the work...what is there to think about?' We do something whilst at work but we do it without interest or motivation.

When our challenges are insufficient for our capability we feel frustrated and switched off. We grow anxious and hesitant. Our self-confidence ebbs away. We lose touch with our intuition and if there are choices to be made they seem obvious and self-evident, tedious and demoralizing. A 'challenge' that requires no judgement is no challenge and no fun.

Potential risks of under-utilisation can include boredom, leading to loss of respect for the work (and sometimes the people who do it). Attention wanders, leading to errors, automatic solutions and erratic performance. Critically, important changes in work are missed as the incumbent distorts the original job, e.g. by accumulating added responsibilities, by expanding lateral boundaries, by taking part of the manager's job or by neglecting the least complex work. In other words, they redraw the boundaries of the job and not always in a way that is conducive to high performance.

Transition

Through using a Work Levels model of complexity we can have a new perspective on transitions.

We all experience transitions in our lives. These are times when we move into significantly new and different situations that we have not previously experienced. These include moves from school to college or into an organization. They include marriage and the starting of a family. They can also encompass divorce or the loss of a loved one. In a work context they include such periods as taking on a first supervisory position or moving from an operational role into a strategic one.

Periods of significant change demand that we alter our prevailing model, gain a new perspective on the world, reach out for understanding, discern patterns and then make sensible judgements in the light of the new model. Our mental-models and the associated breadth and depth of perspective set the conditions for feeling in-tune with the work undertaken and ensure a successful transition to new levels of challenge.

A Work Levels transition is a period in time when a person is tuning into and becoming accustomed to working at a level of complexity that is a step-change from their previous experience. Transitions are positive, but often turbulent times in our development. We have to learn to leave behind some of the approaches with which we have been most comfortable and reach out towards new ways of thinking about the world. They can be likened to paddling across an unexplored and wide river. One has to leave behind the known and familiar place before the destination can come into focus. But if the water becomes choppy it is easy to think about turning back.

Individuals are affected by transitions in different ways. Some thrive; others can become almost overwhelmed with uncertainty and self-doubt. But what is common is that at the end of a transition the individual has the ability to handle more complexity and make those decisions called for by a higher level of work. Transitions are times of opportunity, but also times of added risk.

The drivers of past success will not guarantee future success. More of the same will not be enough to succeed in our next role and through the next phase of our development. Success requires us to abandon previous skills, operating styles and mindsets, not just acquire new skills.

There are two main phases in a transition, namely preparation and actually taking on the challenge.

Preparation is the phase where a person tunes into the demands of a completely different level of work challenge. It is here that involvement in the work of the new level, e.g. via working parties or study groups, special projects etc. that real benefits can be gained before taking on a full role in the new level.

When a person takes on a new work challenge at another level there are certain dangers. They need to beware of over-reaching themselves before their skills have caught up. It is easy to over-estimate one's own current capability. By way of analogy, it is like driving a car at night. We can see the road ahead but in reality our contact with the road is some way back.

A major transition that some people have to make is a move from operational to strategic work. This can be difficult with its greater levels of uncertainty and ambiguity. This is especially so for those trained in and rewarded for excellent logical analysis such as engineers and scientists. Rather than having to let go of all that is held dear, it is necessary to begin to see the limitations of a set of skills that enables one to converge on an answer and to add a range of

107

other thinking skills that will enable new possibilities to be opened up. Then there are moves to take on the top jobs. Receiving the call to the top office as CEO or the head of a Business Unit or large function requires special skills to influence decision-makers and allies inside and outside the organization and requires the individual to acknowledge a much more collegiate decision-making environment, yet they have to be fiercely competitive for resources. This apparent dichotomy takes some getting used to and new entrants often benefit from having a mentor with whom they can discuss and explore ways of dealing with the subtleties freshly encountered.

More than anyone else perhaps, the holders of top jobs are under observation 24/7 and any quirk or surprise is quickly interpreted and invested with meanings. As a top jobholder, the individual is able to exploit this visibility by acting in ways that surprise and which send new signals or symbols into the internal or external environment.

Summary

The Copernican Revolution took place over many years and had to overcome huge resistance. Copernicus' ideas just didn't fit with our perception of the facts. Of course the sun revolves around the earth, it rises in the east, *travels* across the sky and sets in the west! Similarly, in psychotherapy, counselling and coaching, the change is slow and there is much resistance. The fact that humans are contingent and codependent doesn't fit with our perception that the world revolves around us.

To enable the equivalent of the Copernican Revolution, psychotherapy, counselling and coaching in their separate ways have to become ecological studies of human *beings* in their natural habitat and which take the entire context of human living into account in a dynamic fashion.

Chapter 16
Economics Reconsidered

Our experience of life tells us that people are imperfect. Sometimes we're rational and make decisions on the basis of being well informed. Sometimes we act and have no idea where the impetus came from. We only need to stop for a second to think how our behaviour changes when confronted by the word 'free'. Zero is not another price. It turns out to be a hot button connected by a neurological expressway to our unpredictable side. Would you buy something when it is discounted? Quite possibly; if you really wanted it. Would you grab if it were free? Yes, definitely; whatever it is.

Despite this level of self-understanding, policy-makers, and those who studied our economy, made the assumption that people were well informed and rational. When confronted with irrational behaviour such as how we go on gambling bigger and bigger sums to avoid losses, some of the brightest minds have turned this around to show that this is actually rational behaviour because it is consistent.

This theory of rational choice based on ideas of consistency builds on the traditions of thought that came from the Enlightenment conception of 'rational man' that arose in the eighteenth century.

In the 1960s and 70s Milton Friedman and Robert Lucas were thought leaders who built on these ideas of rational man. They developed the crucial insight that the development of the economy depends to a large extent on what people expect. Continuing with this theme, in 1976 Robert Lucas argued that 'any change in policy will systematically alter the structure of econometric models'. In *other* words, policy affects behaviour, so in changing policy any model must take into account expectations.

The views expressed by Friedman and Lucas chime with my assertion that the future is imagined and how we imagine the future depends to a large extent on the system in operation. Indeed, if the implications of this were fully taken on board economists' ambition to predict and control in a complex, contingent world would have been curtailed. However, Friedman and Lucas turned in a different direction and to make their ideas tractable they made the assumption that rational people have rational expectations.

Instead of shining a light on the frailties of our understanding as it should have done, the 'rational people' argument drove economists and econometricians into a navel gazing, mathematical spiral. Rational expectations became the norm and anyone who questioned the assumption of rationality was accused of being irrational themselves.

Whole theories in economics have been built on the understanding that people are rational and this had fed through to social policy-making. As interpreted by the important Chicago school of economics, faith in human rationality is closely linked to an ideology in which it is unnecessary and even immoral to protect people from their choices. Milton Friedman, the leading figure in the Chicago school, expressed a view in of one of his popular books, *Free to Choose* that rational people should be capable of taking care of themselves.

The assumption that individual agents are rational provides the intellectual foundation for the libertarian approach to public policy; do not interfere with the individual's right to choose unless the choice harms others. For the economists who favour this view, rationality is not about what is reasonable, it is about whether there is internal consistency or not. A famous example of the Chicago approach is entitled, 'A Theory of Rational Addiction' (Becker and Murphy, 1988); it explains how a rational agent with a strong preference for intense and immediate gratification may make the rational decision to accept future addiction as a consequence. It seems that if we are predictably irrational then the theory that we are rational beings still stands!

In this world of rational agents (let's call them Econs) governments should keep out of the way allowing us to act as we choose so long as we do not hurt others. If a motorcyclist decides to ride a motorcycle without a helmet a libertarian supporting this school of thought will defend their right to do so. However, there is a hard edge to this. A person who did not save enough for their old age is likely to be given as much sympathy as the person who complains about the bill for the meal at the end of a long and liquid lunch. That way of thinking could get a lot tougher still and can be illustrated by referencing obesity. Obesity is a growing issue (no pun intended) and there is rightful concern for the health and well being of the seriously overweight. However, obesity is an issue that certain libertarians could take a different view on to one of care and concern. Healthcare costs could be exorbitant if you have *chosen* to be grossly overweight! Another pernicious implication of the rational agent model in its extreme form is that customers are assumed to need no protection beyond ensuring that the relevant information is disclosed; an Econ can deal with the small print when it matters.

The model of human nature on which the Chicago school base their ideas is flawed. Humans are not wholly rational beings and are rarely consistent. We have the capacity to want two things at once that are incompatible. Some of us want to be free and independent *and* at the same time secure, loved and completely cared for. Some of us want to be here *and* at the same time there. Humans are not Econs and are susceptible to a whole range of bias that we cannot consistently avoid. By claiming that people are not Econs, it does not logically follow that we have to have a system that chooses for us. Freedom is not a contested value. It is just that freedom is more complex than the Chicago school and those politicians and policy makers who believe in people as rational free agents currently appreciate.

If you think that all this highfalutin academic stuff is just for the birds you could not be more mistaken. The central assumption of perfect rationality ran through the Greenspan doctrine and directly affects whether you are reading this with millions in the bank or constantly worrying about how to pay off the credit card (depending on which side of the hedge (fund) you sat). Alan Greenspan, the then Federal Reserve Chairman, gave a speech in 1999 that laid out his philosophy and thinking on bubbles that effectively defined the economic policy landscape for a decade. Greenspan argued that bubbles couldn't be easily spotted as they form and even if they were, they couldn't be stopped from inflating without causing a substantial contraction of the wider economy. The solution, according to Greenspan, 'was to mitigate the fallout when it occurs and hopefully ease the transition to the next expansion'; an approach that was known as mopping up.

Greenspan didn't have to wait long to test the theory; the bursting of the dotcom bubble enabled Greenspan to mop up by reducing interest rates to historic lows. What he had not foreseen was the effect of creating a debt-fuelled boom.

Even surging asset prices didn't faze Greenspan or the regulators in other countries. It had become the orthodoxy to believe that the only valid valuation of an asset was the market price for that asset. The decades old principle that a valuation had to be a 'true and fair' reflection of the asset was downgraded. However, the market prices of assets do not behave as for other normal goods. The theory that only the market price matters depends crucially on the validity of the efficient markets hypothesis. The efficient market hypothesis was put forward by Eugene Fama in 1970. In this, the market price for an asset is the best possible predictor of the price in the future because all information is distilled into that price. The hypothesis means that no asset can be fundamentally valued wrongly because the market has processed all the known information and the asset will trade at fair value. In addition, no one can beat the market unless they have insider information. The theory is based on ideas of rational man and, not only that, it assumes that rational man is fully informed and that risks will be taken by people best able to bear them.

A common assumption was that, based on the efficient markets hypothesis, the movement of market prices followed a normal distribution. The special property of normal distribution is that you can use it to work out the sort of results that you would expect 95 percent or 99 percent of the time. However, the predicted outcomes were based on crucial assumptions that were nearly always wrong.

111

A set of outcomes can only be described as a normal distribution if what you are measuring is the product of random and independent causes. And we have repeatedly seen people do not behave rationally and events are not independent but contingent. Empirical evidence of how markets move shows how price changes do not follow normal distributions but have the chance of extreme outlier effects.

For an economist, mathematical models should be a form of metaphor. For an investment banker they should be as pale reflections of reality. Yet all too frequently they became simplifications through which economists and bankers blinded themselves to much of the reality that didn't fit the model. There is more on this in the next chapter.

The prevailing model of rational man and the computer provided us with a false sense of control. If you can predict a risk, you can hedge it. If you can find a buyer of that risk you can price it and telecommunications brought buyers and sellers easily together facilitating the growth of the derivatives market. So we had reached what was called a period of stability, no more boom and bust; if everyone knew what the price was at a given point and the risks were aligned to those who could bear them - what could possibly go wrong?

The assumptions about risk were wrong and the unrestrained bubble was of historic proportions. The fallout has been massive at a societal, organizational and individual level. As for the impact on economists, they are likely to be just behind bankers in a popularity contest. It is clear, to me at least, that if economists are ever to fully recover their status, they need to bring time, uncertainty and the nature of *being* centre-stage.

Economists need to have a greater awareness that our 'cognitive spectacles' are heavily coloured by the metaphors we employ and that the models we use about people and markets are incomplete. More importantly, economists need to fully appreciate that in an open system where nothing ever quite repeats and where there is creative choice and large degrees of freedom, economic expectations cannot be purely the product of reason. Your decisions must also be based on how you imagine the future and how you will it to be.

Recasting economics in this way may make the subject seem not quite so scientific but it has the major advantage of underpinning the discipline with a real understanding of the basis on which people actually behave in the real world.

Chapter 17
A Machine for Telling the Future

The ancient Chinese used to consult the *I Ching,* reputed to be the oldest book of divination. It sets out the universal and necessary pattern of change, the very Laws of Heaven to which mankind will conform freely as a result of insight gained from wisdom or from suffering. The *I Ching* is based on the conviction that whilst everything changes all the time, change itself is unchanging and conforms to certain ascertainable metaphysical laws.

The Bible also talks of time and change. Ecclesiastes; 'To everything there is a season and a time to every purpose under heaven...a time to break down and a time to build up... a time to cast away stones and a time to gather stones together' or as we say, a time for hiring and a time for firing.

It is clear that the task of the wise man has been and is still to understand the great rhythms of the Universe and to gear in with them. If not rhythms as such, in many ways it looks like mankind is tuning into the laws of the universe.

A defining characteristic of human progress is our ever-increasing ability to understand, predict and control our environment. Our lives today are what they are, better or worse (largely better I suggest), because we understand much of how the world works. Teasing apart underlying cause and effect relationships in the physical world has not only satisfied our curiosity, it has given us the power to predict the outcomes that matter. As a consequence, much of what once appeared to be in the province of the Gods has been dragged into the realm of mere mortals.

If some predictions prove to be less than accurate, our collective belief is that anything beyond our grasp is only temporarily so and it is only a matter of time before this will fall under our dominion if we persevere in the belief that it shall.

Such optimism is not baseless especially when we are predicting the outcome of physical events but our interest now includes predicting the outcome of man-made entities such as the stock market.

Despite the theoretical difficulty and practical challenges associated with accurate prediction, there have never been so many futurologists, planners, forecasters and model-builders as today and the computer has become the modern 'crystal ball' that foretells the future.

The ancient Chinese went to the *I Ching*, we turn to the computer.

Machines that tell the future are based on a specific assumption that we have already found to be false, i.e. 'the future is already known so that it merely requires good instruction to get it into focus and make it visible'. This is in fact an extraordinary assumption and goes against all direct personal experience. It implies that human freedom does not exist or in any case that it cannot alter the predetermined course of events.

Planners of course proceed on the assumption that the future is not already determined and that their plans will make the future different from what it would have been had there been no plan. Yet it is the planners, more than anyone else, who would like nothing better than have a machine to tell the future.

Now there are those who argue that computers can best predict the future due to their capability of handling vast amounts data that fits to a mathematical expression. By such means it is possible to have accurate and up-to-date information and once you have a good mathematical fit, the machine can predict the future.

This sort of thinking led to the establishment of Long Term Capital Management (LTCM) and its collapse in 1998 when it lost over $5 billion in funds. This fund was set up by the esteemed Nobel Prize winning economists who had themselves created a model for working out the value of complex derivatives known as the Black-Sholes model. Since it was first published in 1976 it became the industry standard and virtually all traders used the Black-Sholes Model to value their books.

Scholes joined fellow Nobel laureate Robert Merton and John Meriwether from Solomon Brothers to set up a hedge fund to make money from the renowned risk management models. The model told them where to invest their money. Specifically they assumed that the US and Russian government bonds would be correlated and move together. They made huge returns by observing movements in the respective markets, selling US bonds and buying Russian bonds at the right time and hedging their risk in derivatives.

Yet, for the all the brainpower in the Company, they made a fundamental mistake in assuming that a 'good mathematical fit' equated to reality. However, 'good mathematical fit' simply means that a sequence of quantitative changes in the past has been elegantly described in precise mathematical language. But the fact that a Nobel laureate has been able to describe this sequence so exactly by no means establishes a presumption that the pattern will continue. It could continue if and only if a) there were no human freedom and b) there were no possibility of any changes in the causes that have given rise to the observed pattern.

When the market moved massively against the direction of the good mathematical fit LTCM took too long to respond and the Federal Reserve had to organize a bailout that involved unwinding all the trades and the investors losing everything.

I am sure that Scholes and the others had a precise appreciation of the assumptions on which they based their model. I am also sure that, over time, they became impressed by the thoroughness of the job done, by the fact that everything seems to add up, etc. Certainly, the person who used the forecast probably had no idea at all that the whole edifice was to stand or fall on the basis of one single unverifiable assumption.

The founders and the users of the models in LTCM failed to take into account what happens when a crisis occurs. The models assumed that markets worked smoothly and that governments were always in a position to pay their debts. The models failed to reflect how correlations change, how asset prices can move suddenly, how liquidity can disappear and how the herd can shift.

If LTCM had presented their working artlessly on the back of an envelope, there would have been a better appreciation of their tenuous character. Of course that wouldn't have made the founders a fortune but it would have prevented their investors losing their shirts in the process.

What is surprising is that this episode didn't warn people of putting too much faith in the machine. It just redoubled their efforts to take more factors into consideration. They would have been better served by stopping and thinking about the nature of time, complexity and uncertainty.

Understanding the nature of time, complexity and uncertainty is not just of intellectual interest. It is paramount when predicting the future. Given that there have never been as many futurologists, insights provided here should be of interest to a broad spectrum. However, I suspect that futurologists might not like what I have to say and, as such, they will ignore the view, deny the validity and defend their position.

Predictability represents one of the most important and practical problems with which we are faced. There are practical and theoretical problems associated with this.

In practice, all prediction is simply an extrapolation modified by known plans. But how do you extrapolate? How many years do you go back? Assuming that there is a record of growth, what precisely do you extrapolate - the average rate of growth, or the increase in rate of growth and the annual increment in absolute terms? It is good to know of all the different possibilities of using the same time series for extrapolations with very different results. Such knowledge will prevent us from putting undue faith in any extrapolation.

The theoretical problems are much thornier. In identifying the limits of prediction, it is worth asking the question 'is prediction or forecasting possible at all'? The future is imagined. How could there be knowledge about something that is non-existent?

In the strict sense, knowledge can only be about the past. The future is always in the making but it is being made *largely* out of existing material about which a great deal can be known. Therefore the basis for sound forecasting is laid, is it not? However, before we get carried away, prediction only works accurately in certain prescribed situations. If we want to know the position of Neptune at the end of next month our prediction would be on solid ground as our ability to predict planetary motion has been finely tuned and proven reliable. It is based on a theoretically sound understanding of cause and effect and has been empirically validated through repeated observation. However, these are not the kinds of facts that form the day-to-day activity of strategists, planners and forecasters. Instead, in business successful commitments must align with variables such as industry forces (including new entrants), key trends (including socio-economic) and macro-economic forces (including the state of the capital markets). In the absence of reliable and sufficiently accurate predictions of these and many other variables making commitments today that pay off if, and only if, they fit with predicted future circumstances demands not just forecasting but also the art of communing with the Gods. Such gifts are in short supply. Winston Churchill summed up his views in 'all prognosticators are bloody fools'.

Of course mere assertion, however forceful, is not proof that forecasting does not work. Given that the forecasting industry includes professional bodies covering such as meteorology, economics, investment, technology assessment and human resource planning it must be the case that some have found the secret to good forecasting and prediction – mustn't they? However, as is very evident in the history of the investment industry it would appear that for all the effort put into peering over the horizon, all we have to show for it is, if anything, collectively worse than my own dismal track record.

If there remains a flourishing, multi-million dollar market for prediction it is, as Samuel Johnson observed of second marriages, a testament of hope over experience. Michael Raynor

Prediction is heavily influenced by the nature of time, randomness and free will.

Proponents of prediction and forecasting will point to forecasters who routinely get it right. Within the finance industry fortunes are paid to those, sometimes armed with mighty computers and sophisticated models, who claim to call the forward LIBOR or tell you which sector of the stock market will outperform this year. However, these predictors are more prophesiers than forecasters. Forecasting, to be of any validity, has to be wrapped in the appropriate probabilistic hedges. That is because the future is imagined; it is here now along with our history and our experience of the current moment. There is no specific future out there waiting to happen as time's arrow carries us towards it.

When viewed from the perspective of today, there is a probability of any given event occurring at any given point in the future. Summing across all events, all time horizons and all probabilities we can conceptualise a 'possibility space' encompassing all events and their associated likelihoods for all times.

In contrast, forecasts as generally used for planning tend to take the form of 'next year's sales will be $100m'. This is nothing more than a guess. A guess that turns out to be correct is nothing more than a lucky one.

The problem with forecasting gets worse. Even if we were to couch all predictions in probabilistic terms we still would not know how accurate our predictions have been. For example how would one measure the accuracy of the prediction that there is a 15 percent chance that the stock market will go up to tomorrow? Since we cannot rerun the events from yesterday to today as many variables have changed, we cannot determine if that prediction would be right 15% of the time.

If talking of probabilities is bad enough, two facts put predicting accurately and usefully enough for strategy-making permanently beyond our reach, namely randomness and free will.

In general usage 'random' typically means without apparent order or pattern. What is meant by pattern needs to be carefully understood for, as we covered in Chapter 10, humans have a propensity to identify patterns even in random data. If the winning lottery numbers are 1,2,3,4,5 and 6 someone might conclude that these numbers are not random because there is a readily apparent order.

When in basketball a player sinks four or more baskets in a row the inference of a pattern is irresistible; they have acquired a 'hot hand' and other players will pass to the person more and the defence will mark them out. Similarly, in the finance industry, a run of good investment calls will equally endow the person with the title, 'the one to watch'.

When we look at the lottery numbers, for instance, it clear that they are in effect random because they are merely a sampling from a larger list. 1-6 does not represent any underlying order because these numbers were just as likely to come up as any other 6 figure combination and it provides no indication of what next week's numbers will be.

The tendency to see patterns and discount randomness is overwhelming. We are wired to detect patterns and the illusion of patterns affects our whole lives. How many good years should you wait to see if an investment advisor is unusually skilled? How many successful acquisitions should be needed before a board decided that a CEO has extraordinary flair for such deals? How many goals must a talented striker score before they are feted as the salvation of the national team? If we follow our pattern-recognition-led intuition we will decide too quickly and jump to conclusions by misclassifying random events as systematic.

Randomness can come from variables being injected into an otherwise orderly system from its external environment.

If the success of a new product was not all that was planned and predicted it might be because of the unexpected launch of another product that came from left field and included previously unused technology. This kind of randomness can be overcome by extending the boundaries of the system studied. In the case of the new product launch the outcome was unexpected because the boundaries of the system being modelled were not drawn broadly enough. But when do you stop expanding the boundaries? In understanding market evolution, when do you stop including variables that might otherwise affect the outcome? In other words, when randomness can be injected from the environment your only response is to include the environment in your system. This is a slippery slope; when a theory of everything is needed to have a theory of anything the computational complexity becomes overwhelming.

Randomness also comes from the initial condition. A system may be orderly but highly sensitive to its starting position. If the starting position, i.e. the initial condition, is anything less than orderly, a process of amplification transforms those inputs into random output. This is the domain of chaos theory and is often described as the Butterfly Effect. The basic idea is that a butterfly flapping its wings over Thailand could cause a tornado in Texas.

Chaos and complexity theorists shift our perspective from seeing reality as static models and stable patterns to seeing reality as living and evolving systems that account for the phenomena of emergence, evolution, bifurcation, indetermination and flow. According to this view, systems emerge from the bottom up and the parts embody the whole. From this perspective, the vitality of systems is on the border between chaos and order.

The essence of life itself and not just organizational life is the antinomy between chaos and order. Life maintains, but only just maintains, a control over the clashing elements that compose it.

The inescapable conclusion is that randomness, the lack of order or pattern, is a necessary component of every system we might want to understand and control. We are doomed to either draw our boundaries too narrowly, leaving ourselves open to an injection of randomness from the environment or to under-specify the initial conditions that determine the ultimate outcome. Frequently we are victims of both. Either way, it is uncertainty, not predictability, that best characterises our future. Michael Raynor

Randomness also comes from human free will.

It is the intrusion of human freedom and responsibility that makes human affairs largely unpredictable. We obtain predictability of course when we, or others, are working to a plan. But this is so precisely because a plan is the result of an exercise in the freedom of choice. The choice has been made; all alternatives are eliminated. If people stick to a plan, their behaviour is predictable within certain limits simply because they have chosen to surrender their freedom to act otherwise than prescribed in the plan.

Experience shows that when we dealing with large numbers of people many aspects of their behaviour are indeed predictable; at least for a while. Out of a large number at any time the majority are working to some sort of plan. Yet all really important innovations and changes normally start from tiny minorities of people who use their creative freedom to develop new plans or ideas. Change can sometimes cascade through the rest of the population like a thermonuclear reaction. Suddenly, and without warning, the human herd can shift direction.

In the business world change happens through entrepreneurs and entrepreneurs are often unpredictable even to themselves. Indeed, what is true of entrepreneurs is true of the rest of us as individuals. The way that we will respond to a given stimulus is not something that we can predict even of ourselves. Clearly it is not our rational faculty that determines our behaviour and without a moral compass, for many, it is not our moral faculty either. The result is that human behaviour is, deeply and irretrievably, unpredictable. Sometimes we choose to be unpredictable for rational reasons and sometimes we act in an unpredictable way simply because we cannot help ourselves.

Although unpredictability is built into life and especially business life, that hasn't stopped many organizations from growing large head offices to accommodate the need to obtain reliable knowledge about an essentially indeterminate future. In these offices, the action-oriented executives surround themselves with an ever-growing army of forecasters, an ever-growing mountain of factual data to be processed by ever more complex algorithms. The results are little more than a game of make believe and an ever more marvellous vindication of Parkinson's Law.

Work expands so as to fill the time available for its completion. Parkinson's law as first articulated by Cyril Northcote Parkinson.

119

The development of what purports to be better forecasting can become a vice. But what can even sophisticated techniques say about short-term forecasts that crude methods could not? After a year of growth there are only three main forecasts, 1) we have reached a temporary ceiling, 2) growth will continue at the same/faster/slower rate and 3) there will be a decline. The choice between these three basic alternative predictions cannot be made by forecasting techniques but only by informed judgement. Once you have fully understood the current position and identified all the abnormal and non recurrent factors, no amount of refinement will help one come to the fundamental judgement – is next year going to be the same as last year or better or worse?

Crude methods of forecasting (after the current picture has been corrected for abnormalities) are not likely to lead into the errors of spurious verisimilitude. Once you have a formula and a computer, there is an awful temptation to present a picture of the future, which through its very precision carries conviction. Yet a person who uses an imaginary map thinking that it is a true one is likely to be worse off than someone with no map at all; for that person will fail to look at the detail of their position and their surroundings and make the necessary adjustments to keep them on the right path.

It is clear to me that the best decisions will continue to be taken by mature non-electronic brains possessed by people who have looked steadily and calmly at the situation and seen it as a whole. 'Stop, look and listen' is a better motto than 'what does the forecast say'.

This may be a case of 'it's my book and I will moan if I want to' but my experience of raising finance throws some light on the current approach to prediction and associated risk assessment. It seems that investors in start-ups are rarely entrepreneurs in the true sense of risk-takers. They want all the projections (predictions) in the business model to point to a risk-free decision before backing a venture. It is a strange world in which the hubris of banking bosses can lead to the acquisition of vast amounts of toxic debt whilst investment in start-up companies only happens at a point when the owner has been successful and no longer needs the money.

Chapter 18
Lies, Damn Lies and Human Resource Management

The social and organizational systems that underpin the management of people are critical to success. By social and organizational systems I mean those that are controlled by the CEO and include the processes by which people are employed, the pay that is received, the appraisal and recognition that is provided, the career structure and opportunities for advancement and the setting of accountabilities through the vertical and lateral structures.

The systems that the CEO chooses or agrees to have implemented through the specialist HR, Personnel or People function have a huge impact on every single employed person. The net result of the systems is a sense of engagement or alienation.

The sad reality is that the social and organizational systems in operation are, at best, marginally deleterious. At worst, they create huge feelings of resentment, unfairness and distrust.

How come we have produced a deleterious system that is promulgated by the best business schools and reinforced through professional institutes? The clear answer is this; social and organizational systems build-in a range of problems because there is no coherent understanding of the whole person underlying the practice of Human Resource Management. In the absence of this, practitioners have used a range of analogies, substitutions and received wisdom that lacks coherence. This has led to unclear concepts of managerial accountability and authority, no concepts of vertical or lateral organizational structure, dysfunctional processes for the selection and development of talent, overly complex performance management systems and less than fit for purpose compensation systems.

The subject of Human Resource Management is people; how to organize and manage them so that their collective efforts meet the needs of the enterprise. Yet the function applies next to nothing of our understanding about what makes people supremely adaptable or what we know about how people think. Thinking is the essence of being human yet, if it is taken into account, it is limited to the classical aptitudes of logic and analytical reasoning. As a result the people domain is reduced to, at best, a consideration of constituent parts or an aggregation into a collective mass.

It is as if sailors have lost interest in ships.

The analogies, substitutions and received wisdom utilised by HR practitioners include the following:

Let's treat labour as if it is a commodity and like any other commodity establish a bargained price according to market forces.

Commodities do not have a contract with their owner!

Individuals not only have a written contract with their employer they also have a psychological one as well.

Individuals cannot contract or guarantee to produce the results, only to use their best endeavours to do so. This is because firstly, they do not control the resources they have available and secondly, neither they nor anyone else can foresee or control the conditions and circumstances that will arise as they get on with their work.

The best endeavours and values of the individual are waived aside by the inventors of incentive schemes in an attempt to prod people in a particular direction. The end results are less than hoped for as production workers ship poor quality to meet quotas, sales-staff collude with customers to meet sales targets and CEOs manipulate share-prices to maintain management bonus levels.

When we treat labour as a bargained entity we get the collective antagonism that goes with it.

Would we treat labour as a commodity if we fully appreciated the power of the individual once they have contracted to make their best endeavours?

Let's treat people as greedy and lazy and create compensation systems to control this (or in the case of greed - sometimes exploit this).

Compensation practices are of eminent importance for millions of wage/salary earners and their families for compensation determines standard of living, socio-economic status, economic security, self-esteem and the distribution of wealth in capitalist nations. Unfortunately, most compensation systems are based on unsound notions of human nature. The systems that are put in place suffer from a host of misconceptions that include, 1) people are lazy and need to be incentivised by being held accountable for their own results and receiving pay that accords with these results, 2) work is something people are forced into in order to survive, so incentives are essential in getting people to give their best efforts.

These misconceptions force employees to be self-seeking and grasping and to cut corners on quality in order not to lose out in the compensation stakes. Under fairer treatment however, people are capable of behaving fairly, reasonably, honestly and cooperatively even about pay.

Let's treat organizations as if they are machines.

The word that defines our industrialised age is 'machine'. Although the word has been in use for millennia it really came into its own when machines were powered by more than human or animal effort. Steam power, and the sources of energy to come, took the grunt out of repetitive work and enabled the replication, time after time, of components with absolute precision. 'Replicability' heralded mass production.

The machine is almost revered. So when we see an organisation, civilian or military, doing what it does best we say it's a well-oiled machine. This is strange because we don't apply the same machine label to a beehive or ant nest. We don't because the term doesn't apply to organical work.

Given the complexity inherent in life and that there is a social world in which we all are locomoting; all moving to a purpose; all making choices, why are we so happy to pin the machine label on the human-created organization? The one thing about a machine is that it does not choose. Is that really what we want to create?

Would we treat human organizations as if they are machines if we fully appreciated that they are made of individuals undertaking organical work; each making decisions in the absence of complete data.

Let's focus on behaviour although behaviour is an outcome of how we think.

To bring 'science' to the party, many theorists and practitioners have focused on what we do. The resulting competency frameworks solely focus on demonstrable behaviours that are associated with high performance.

Whilst well intentioned, these behaviour-based frameworks actually miss a key point. Behaviour is an output; it is an end result of something else.

Would we focus on behaviour if we fully appreciated that thinking about the future (i.e. our expectations) drives behaviour?

Let's put in complex performance management systems that substitute management judgement for measures of output.

Ask any CEO what their ideal human resources department would be able to deliver and their responses are most certain to include:

- Provide a high performance workforce,
- Equip the line managers with tools to manage their people that are easy to use and really work,
- Develop and implement policies that serve the organization well.

Yet, decades after CEOs have asked for this, few HR departments have really delivered.

Many HR professionals claim to have the answers for providing a high performance workforce, but relatively few organizations have processes in place that really nurture or enhance performance. Research consistently shows that employees are more likely to be more de-motivated than motivated by their performance appraisal systems.

In the course of my career I have introduced a number of performance management systems; all, I have to admit, with a fatal flaw, the same fatal flaw inherent in virtually all systems. I will illustrate with a personal experience.

Once upon a time I was young and enthusiastic and, on joining a new company, keen to impress. At the annual target setting date, I had already recognised where changes needed to be made and I was happy to sign up to a range of 'stretch' targets that would form the basis of my appraisal in a year's time. Having set the targets, the company decided to introduce a total quality programme designed to impact all company processes. Firstly, all existing processes had to be described in detail. Then there was employee training to arrange and constant updates. The year went by in a whirl. It had been a good year, I thought, until the appraisal! Some of my stretch targets had not been achieved and the appraisal became a defensive exercise and one of damage limitation. The result of the year's hard work was an average score, an opportunity to go on a Time Manager course and a wiser, less enthusiastic me when it came to setting targets for the year ahead.

If the fatal flaw is not obvious, let me say it in one sentence. This flaw is a general belief in holding people to account for what they produce even though the manager determines success.

In fact there are other flaws in many other performance management schemes especially those that force the performance distribution. Forcing the distribution means mandating that, say, only 10% of the workforce can be on the top performance mark and 10% must be in the bottom. This sort of scheme presumably is based on the idea that all salmon swim faster if the slowest 10% are culled!

Let's fill up the organization with the best people even though 'best' is completely dependent upon the fit to the job required.

Sometimes a book's title is so good that it sums up the whole book. 'Men are From Mars, Women are From Venus' is one such. In the business realm a good example is, 'War for Talent'. This was written by Michaels, Handfield-Jones and Axelrod, who were part of consulting firm Mckinsey. I wonder how many CEOs or HR Directors actually read this? Definitely not as many as those who were inspired by the title to join the mad scramble for top talent from the best business schools and lock in the 'best' through the ever increasing use of 'golden handcuffs'.

The words 'best' and 'talent' are still bandied around as if they are context-free elements. But talent is only talent if it is relevant for the job required and 'best' is absolutely situationally dependent.

When it comes to thinking about 'best' the only best that is really worth taking seriously is 'best-fit'. However, current systems mitigate against this. Rather than job descriptions and person specification dovetailing together, there is often a systematic disconnect in the ways that jobs and people are described. With the different technologies and methods associated with job evaluation and psychometrics, it is as if jobs are measured in units of time and people in terms of weight. When trying to relate people and jobs using different measurement systems it is like trying to ask, 'how many minutes are there in a kilo?' The consequences of this disconnect are most evident in the area of resourcing. The people and job matching processes including recruitment, selection and the management of talent have become complex, time-consuming, massively costly and inefficient. Not only that, the result of the matching process is often a wrong fit.

Let's measure everything so that we can manage it, but let's gloss over the fact that HR measures fails to provide executives with information on which they can make strategic decisions.

The measurement of human capital promises a great deal. However, despite significant effort, the human resource of a business has defied easy evaluation in the way that financial capital and markets have been assessed. Whilst finances can be talked about in clear and consistent ways, there is no common language that enables Human Capital to be addressed.

The good news is that accounting and management professions recognise that traditional corporate measurement systems must be enhanced to take into account the value-add of the people domain, especially in the knowledge-based economy. As yet though few organisations have the ability to assess or fully utilise this knowledge-based asset, despite many hours of searching for ways of measuring it and managing it.

In Finance and Marketing the frameworks that they use provide line managers with tools that help them focus on operational efficiency and there is a sense of cohesion about the practices that underpin the linking of money to business opportunity and product to market opportunity. This provides the basis to make important decisions in an aligned way. Although managers might not like, for instance, the fact that some areas are allocated more resources than others they understand that investment must flow to the units that are most financially pivotal in helping meet the strategic and operational requirement.

Compare this to the position within the people domain. Despite the investment in sophisticated software systems and balanced scorecards, HR measurement is largely focused on the outcomes of work that is undertaken by the HR function. In the words of John W Boudreau and Peter Ramstad, 'HR measurement is largely produced by HR people for HR people'.

HR will never have measures that are equally significant so long as they focus on the after-event activities or benefits of the HR function or programmes. To be strategically significant they must focus on improving people-related decisions that are made by senior management.

Team working is a good thing, so let's make the team accountable for delivery.

Edgar Schein posits that 'every organized activity gives rise to two fundamental and opposing requirements'. These are namely: the division of labour and the coordination of effort. The division of labour is about chunking work into discrete work packages and the coordination of effort relates to the holding together of related activities to meet the end goal.

Leaders, managers and human resource professionals attempt to address the 'coordination of effort' via the introduction of group dynamics training, the use of group decision-making, self-managed teams and various kinds of group and department bonuses and profit sharing. While these approaches may be effective in the short term, they are not sustainable over time because they are in direct conflict with the contract under which all employees have always worked.

If only there were agreement in the team, wouldn't everyone get on so well and produce wonderful results. This sentiment has resulted in large sums being spent on team events and bonding sessions.

'The key to unlocking openness at work is to teach people to give up having to be in agreement. We think agreement is so important. Who cares? You have to bring paradoxes, conflicts, and dilemmas out into the open, so collectively we can be more intelligent than we can be individually'. Edgar Schein

Let's get rid of the management hierarchy to improve communication, process flow, decision-making and efficiency.

Somehow the word 'hierarchy' has become a dirty word and when associated with 'bureaucracy', well, wash my mouth out! These two 'ills' of modern life are now blamed for a range of communication problems and are synonymous with inefficiency and politicking.

The 'down with the hierarchy' movement fails to recognise that the hierarchy, as a cascade of accountability, is the only structure that truly fits human nature. The hierarchy has naturally evolved with society throughout its 'post-tribal' history and all attempts to establish a sustainable alternative form have failed.

Contrary to popular myths, well constructed hierarchies are the basis for flexible and creative organizations.

Poor organizational structures have two serious consequences. Firstly, because there is a strong tendency to create control systems that reduce managerial judgement as far as possible, there is a serious loss in managerial effectiveness. Secondly, top management wastes time and effort by focusing on training, exhortation and other change efforts as a means of transforming values and attitudes so that people behave and work together differently and more effectively.

In reality the exact opposite of control and exhortation is needed. Rather than changing behaviour and outlook, it is more effective to develop better managerial leadership systems. In other words, systems need to be implemented that take full advantage of the judgement and decision-making capabilities of managers at all levels and provides them space and authority to produce results through their subordinates under their leadership. A clear and well-structured hierarchical organization is the prerequisite.

Let's identify a recipe for leadership success so that we can clone managers.

In less than ten years the world's financial system was subjected to two seismic events. Firstly there was the massive devaluation of all things tech-stock. Then the credit-crunch made the internet-bubble look like the froth on a derivative trader's daily cappuccino.

What can we put this down to? Are there simple explanations? May be? Greedy bankers and brokers promoting over-inflated stocks, the herd instinct, dubious accounting practices and plain foolhardiness are likely candidates to take the rap. However, the humble competency framework is unlikely to be called to account. Yet, it has left a grubby fingerprint at the scene.

The competency movement has been the big management thing over the last three decades. Companies big and small are now creaking under the weight of the frameworks that identify success; especially leadership success. They were supposed to enable the selection and development of leaders who could take us to the Promised Land, not simply replicate faulty thinking that leads to repetition of past failures.

The competency movement arose out of psychology's fixation with behaviour.

Most competency frameworks are predicated on the basis that tomorrow will look like today and very few take into account the fact that uncertainty and unpredictability is the way things are around here. If tomorrow is not the same as yesterday, what do leaders draw upon to help guide them to make wise decisions? Clearly, past experience or yesterday's competency list cannot be the whole answer. One of the reasons for having a competency framework is to help people act in a relevant and appropriate manner. But if appropriate behaviour is conditional upon the complexity of the environment within which you work, certainly the 'one size fits all' competency framework will be sadly lacking.

If organizations don't want to be surprised and overwhelmed by so-called unforeseen change, management theorists and practitioners had better wise-up to the limitation of current competency frameworks. At best, these identify yesterday's high performance factors. At worst, they are a mish-mash of behaviours, values and aspirations.

Many leadership competency frameworks tend to establish false and disruptive formulations of the nature of leadership by treating it as a set of essential personality characteristics that have to be cultivated and selected for. We create endless trouble for ourselves by separating leadership from management and should recognise that leadership is about the outputs that make a difference. It is about the decisions that move a unit forward and as such leadership is an accountability that is part of ordinary effective management.

Let's focus on people's strengths whilst ignoring the fact that these can become a person's biggest weaknesses.

Marcus Buckingham in *The One Thing You Need to Know* outlines an over-arching life strategy that he believes will out perform any other approach. It goes like this; understand your main strength and capitalise on that by making it even stronger and utilising it as much as possible. Everyone should focus on their strengths not weaknesses. This is a compelling message that incorporates an important truth. We gain important leverage in life by knowing what we are truly good at and developing that talent to turn it into a source of excellence. However, in the process of simplification, the message of 'focus on your strengths' glosses over the complexity of human behaviour and the reality that strengths can be weaknesses.

The fact that strengths can become weaknesses is apparent when you consider the flip-side of common attributes:

> He is a **team player** might also mean that he is not a risk taker, is indecisive and lacks independent judgement,
>
> She is **customer-focused** might mean that she can only focus on what the customer wants now and might miss future opportunities,
>
> He is **action-oriented** might also mean that he can be reckless

128

and dictatorial,

She is a **people-person** may suggest that she is soft and cannot take tough decisions.

Beware of the strengths! A well-known assessment company prides itself on identifying the strength 'spikes' in an executive's profile. They consider these as part of their winning formula. Whilst spikes might set someone apart and have contributed to their success to-date these spikes need to be treated with utmost caution.

On the basis that our strengths can be our limitations if overused or used inappropriately, there can be a 'darker' side to the very aspects that appear to be part of winning formulas such as:

Alertness............... Narrow focus
Charm.................... Manipulation
Confidence.............. Sense of infallibility
Control.................... Inflexibility
Commitment............ Blind faith
Courage.................. Foolhardiness
Dedication............... Workaholic
Forthright............... Insensitive
Quickness............... Over-hastiness
Sharp wit................ Cutting or Abrasiveness
Perseverance........... Resistance to change
Thriftiness.............. False economy

Especially under conditions of stress, the positive 'spikes' can be over-worked. What used to be a winning approach reveals a fatal flaw leading to derailment.

'Men fall from great fortune because of the same short-comings that lead to their rise' Jean de La Bruyere

Chapter 19
The Radical Alternative

There are calls from many for HR to concentrate on getting the basics right. Paying people on time is a basic requirement, yet this service can be provided through the finance function. Providing clear contracts of employment is a basic requirement, yet this service can be provided through the legal department or the company secretary. Recruitment is a basic requirement, yet is often done best by the line manager. Clearly, getting the basics right can be achieved by eradicating the cost of providing an HR department.

Eradicating the HR function is one thing but the social and organisation systems including organizational structure, selection, career development, personal effectiveness appraisal, merit reviews, compensation systems, job evaluation and assessment that build mutual trust are good for efficiency, good for people, and good for the nation. So, before the HR baby is tipped out with the bathwater, the discipline needs to utilise the scientific basis upon which its strategic contribution can be based.

The radical alternative offered here contrasts with the normal approach of management systems as they attempt to reduce uncertainty through structures and control system and after-the-fact measurements. Uncertainty must not be dealt with as an afterthought. Instead, uncertainty must be placed at the heart management theory. It is only when uncertainty is at the core that organizations can cope with the change and turbulence that characterises modern life.

The radical alternative brings uncertainty centre-stage but it doesn't advocate throwing people into confusion and chaos. It provides a system that brings out the best in people and enables them to go beyond the limited constraints of specified processes. It encourages them to make decisions within their level of capability.

The radical alternative throws out management speak, psycho-babble and HR ephemera and is underpinned with a belief; when jobs have clearly defined accountabilities; when people are capable of undertaking these; when they want to do so and are given the appropriate space to do so, human endeavour can produce the most remarkable achievements.

'The most far-reaching, dramatic and rapid changes in the behaviours of individuals in organisations can be achieved by changes in organisation structures, the setting of appropriate accountabilities and creating the right space in which people can express rational and *trustful behaviour.' Elliott Jaques*

Identify the Value System

Many years ago Henry Ford asked, 'Why is it that I always have to have a whole person when all I want is a pair of hands?' That, I feel, summarised his value system in a nutshell.

There is always a consequence to everything. If you treat people as a pair of hands you will get a pair of hands but not the brain that tells the hands to help make something a bit better than planned.

Henry Ford's question could now be updated to, 'Why is it that I always have to have a whole person when all I want is a brain?'

If you treat people as a brain, you will get very clever outputs with little connection to the real world of making incremental shifts through hard-earned trial and error.

Are we any different from Henry? Do we really want the whole person? Would we really not want to utilise a person's specific skill, knowledge or expertise and disregard the rest? It might make management easier if we could pick and choose the elements that we really want. Yet, people come as a whole package. As such, managing people is entirely different from managing perishable goods or commodities or other inanimate resources.

Of course, the employment relationship is based on a trade of skills, knowledge and expertise for compensation but if that is all you are prepared to accept and disregard the rest there is a slippery slope to using people as resources that you switch off and on again like electrical apparatus. The only difference is that this apparatus can choose to be switched on or not!

Build the Psychological Contract

A contract makes things clear; who owns what, what will happen in terms of fulfilling the contract and what are the consequences of breaking it.

In the case of employment, the contract goes much further than the terms and conditions set out in an offer letter.

The contract is individual and personal. People are not employed as groups and cannot be held responsible for working as an accountable group; only the board of directors falls into that category. Despite trying to dilute this through collective agreements, the contract is essentially between the employer and the individual rather than the group. The effective way to gain social cohesion and positive collaboration within managerial organizations is to make as explicit as possible the individual basis of the contract.

Employees demonstrate a great sense of reality about what a contract of employment extends to without necessarily being able to readily formulate this and to express it in language. The sense of reality has to do with the fact that an employment contract is one in which employees agree to commit themselves to use their full potential capability and skilled knowledge to endeavour to produce the results they are assigned within quality standards and target times.

In establishing the basis for sound contracts it is necessary to understand that, and here is a really tough thing to get your head around, it is a property of managerial systems that managers, not subordinates, are always accountable for results. This goes against most of what is assumed to be good management practice. The ideas stem directly from the writing of Elliott Jaques. When I first came across his ideas I reacted against them. On reflection though, they have a deep vein of common sense running through them. What this perspective makes clear is that the salesperson is not accountable for results, the production worker is not accountable for outputs, and the engineer is not accountable for design quality. **Individual employees can only be held accountable for their full commitment to carry out assigned tasks.**

As part of the contract, managers must be able to assume that, as a matter of the employee contract, their subordinates are always doing their best, and, if this is the case, there is nothing more they can do. Subordinates encounter many unpredictable circumstances, such as flaws in raw materials and component supplies and/or unexpected moves by the competition, which are outside the control of the individual employee. Thus, it is up to CEOs and managers to deal with such disrupting variables by modifying assignments/methods, making adjustments to cope with unanticipated changes in external conditions, adding resources, giving more time, coaching subordinates and/or reassigning work.

The employee contract is about agreeing to pay employees a wage/salary for doing their best, or using their full knowledge and best endeavours to carry out assignments. Although they may not always succeed in achieving a specified goal, they must be able to show that they tried their best to do so.

Clarify Accountability and Authority

Lack of clarity and incorrect assumptions about managerial accountability and authority are the starting point for dysfunctional people systems.

Accountability and authority are at the heart of all interpersonal relationships. These factors establish where people stand in relationship to each other, determine who can say what to whom, who must say what to whom and establish who can get whom to do what. As such managerial hierarchies gain their success or failure through the success and failure of the working relationships between and

133

among their people.

Key to the success of working relationships is the job description.

Making the humble job description such an important element or success may be a surprise to some as there has come to be a received wisdom throughout organizations that job descriptions and person specifications are 'old technologies' that do not take into account the rapid pace of change in modern working life. Given the way that these documents were written as a list of duties may be this is understandable but, in reality, well-written descriptions are the building blocks of success.

Imagine what the world's tallest building would be without the architect fully specifying the components.

In any organization the specification of jobs is essential to both the organization and the individuals working in it. The job description is the formal written document that defines the jobholder's responsibility to operate within his or her job boundaries and defines what specifically the jobholder can be held accountable for. Accountabilities are the critical component. Allocating these clearly without gaps and overlaps is the foundation of high performance.

It is crucial that CEOs model accountability and authority with their senior executive subordinates and require them to do the same with their subordinates.

If a line manager is to be held to account for the output of his team then the manager needs a minimum set of authorities to enable him/her to ensure that direct reports carry out the work that is assigned to them. The line manager is the 'boss', who is individually accountable for the performance of direct reports. This managerial authority is summarised by the mnemonic **VAR4**.

- **V**eto; line managers must have the last call on who works in their area,
- **A**ssignment of tasks; line managers must be able to assign tasks to those who work for them,
- **R**
  1. **R**ecognition of performance; line managers must give visibility of performance,
  2. **R**eview work undertaken; line managers must be able to review the work and,
  3. **R**eward differently; line managers must be able to provide for different levels of applied skills, knowledge and expertise,
  4. **R**emoval; line managers must have the authority to initiate the removal of an under-performing team member.

Match People and Jobs

Accountability and authority cannot exist effectively if the CEO does not have specific and accurate information about the size of the roles in the organization, the size they should be in order to get the work done and the level of capability (the 'size) subordinates must have in order to fill these roles and their actual level. Job evaluations, thought to be objective measures because numbers are used to express the final results of various evaluation processes, are inept measures of job size. Similarly, IQ testing, Myers-Briggs assessments and other types of intelligence and personality tests also fail to give an objective measure of the size of a person, especially as it relates to the size required for a given level of work in a role.

In striving for efficiency and cost reduction, HR functions need to recognise the basis on which high performance is established, namely; by creating a dynamic link between jobs and people.

To make the dynamic link between jobs and people it is necessary to take into account the complexity of the decision-making environment. Once this is established, it not only clarifies the types of decisions that need to be made and the outputs of these, it also identifies the thinking skills and mental flexibility needed to operate successfully within a given context. Once thinking and mental flexibility are clearly defined they become vital ingredients in identifying a person's potential to work at a level they have yet to experience and for assessing the match to the work undertaken.

Manage Performance

The measurement of achievements has become the mantra of performance management. In itself measurement is essential but if you are being held to account for what your manager should be responsible for then the scoring of achievement and translating this into monetary reward will be counter-productive.

Complex performance management arrangements can be removed without fear of impacting performance if they are replaced with a system that makes the manager responsible for the output of their people and enables the individual manager to make judgements about a subordinate's personal effectiveness in achieving the results.

Targets and performance standards are highly relevant as people need to know what they have to produce and by when. However, when it comes to appraising performance it is much better to measure people on how they went about their work rather than what they produced. When equipped with useful frameworks that help managers identify the skills, knowledge and experience that enable high performance, the appraisal meeting can move from being a defensive discussion to an open, developmental one.

If the HR system demands a score, then this does not have to take months of discussion. All managers are able to discriminate between the personal effectiveness of their subordinates. However, it is as if all discrimination (and not just that based on race, creed or colour) is to be avoided. As a result managerial judgement and perception is replaced with overly complex performance measures. Yet, with the manager's manager acting as an important check and balance, managerial discrimination based on their perception is all that is required.

This method of appraisal highlights the distinction between performance (what) and personal effectiveness (how). Performance defines the subordinate's results that can be counted objectively but which the manager should be held to account for. Personal effectiveness, on the other hand, defines how well a subordinate is judged to have done in achieving those results. This distinction is of value because, as regards performance, it sharpens people's ability to do their day-to-day work more successfully and allows them to implement an early warning system. When a person's performance is not being measured in terms of results that they produce, subordinates feel they can freely approach their managers when they find it impossible to achieve the assigned results.

Manage 'Lateral Stretch'

People management is a critical aspect in high performance organizations and time is taken to ensure that managers are not overwhelmed by their Span of Control.

Span of Control refers to the number of direct reports who share the time and attention of a manager and look to him/her for leadership and coaching.

Many textbooks will tell you that to be effective a manager must have between 3 and 8 direct reports. Any more and they lose control. Such ideas on spans of control have allowed the building of empires through the introduction of 'span breakers' and have no basis in theory or fact. In reality, a manager may have up to as many immediate direct reports as the operating circumstances require. One of the fundamental limitations is the need to have appraisal-related discussions. When managers are freed from measuring every aspect of output and are allowed to appraise people on the basis of their perception of a subordinate's personal effectiveness, the greater can be the Span of Control.

Create Effective Compensation Systems

The loudest message an organization can give about what it values most - is to pay for it. Yet in many organizations with their complex mix of base pay, short-term bonus, long-term incentive plans, deferred pay including pensions and benefits the message is often drowned out by the noise of the system.

It only takes a very simple compensation system to elicit a

performance culture. This involves creating a pay structure that provides an equal pay range for roles of equal contribution regardless of occupation; the higher the level of complexity, the higher the level of contribution and the higher the pay. It also involves holding individuals accountable for being fully committed and paying them within their range at the step that matches their manager's evaluation of their effectiveness.

When contribution is the basis for reward it recognises that internal equity is just as important as market competitiveness.

Bonus schemes should come with a health warning and should only be implemented when all the side-effects and possible unintended consequences have been identified. The use of such schemes implies that you are receiving a base salary/wage for less than full commitment and their operation often undermines managers by handing some of their authority over to an impersonal and dysfunctional payment mechanism. This reduces management accountability and it weakens the relationship between manager and subordinate.

Implement Healthy Organization Structures

Based on the research of Jaques, the Levels of Work referenced in the taxonomy have great significance for the hierarchical organization structure. Anyone working in a level in which their boss is operating will feel as though the manager adds little value to his or her work and does not have the necessary vision to set the context. Any manager with an immediate subordinate whose role is farther down than the next level will feel 'pulled down into the weeds,' managing tasks that are too short for the manager's role. Finally, any subordinate with a manager whose role is in the next higher level will feel just the right distance away.

Work Levels can also be used to discover the fundamental structure for effective managerial hierarchies. When this structure is applied, it provides the ideal way of translating strategy into operational excellence and the right space for people to make their best contribution.

Establishing the right number of layers of management is the first step in building an effective organization. Thus, any CEO who wants to establish an effective managerial leadership system will begin by establishing a structural foundation levelled in accord with the time-span framework. All that needs to be done is to measure the complexity of the CEO's role, place it in its proper level and count down from there to determine the number of management layers the organization should have.

Measure Only What is Useful

137

CEOs have become infected with the common, yet destructive, assumption that certain qualities, personality traits, and/or competencies are required for certain types of work. However, there is no such thing as an ideal personality for any role. Successful sales people include introverts and extroverts and leaders similarly come in all shapes and sizes. Everyone needs to possess reasonable amounts (not too much and not too little) of sociability, initiative, aggressiveness, optimism, realism, risk taking, honesty, intuition, loyalty, reliability, balance, cooperativeness, etc., in order to work effectively in any position.

Once it is recognised that the employment contract implies that reasonable behaviour is required, a much simpler and much more valid approach to the evaluation of human capability becomes available. With this method, CEOs will get what they really need; the ability to ensure the placement of the right people, in the right roles, at the right time, at all levels and in all functions in the organization.

The approach begins with the understanding that there are only three qualities that individuals need in order to be able to function successfully in any given role in a managerial hierarchy:

    1. The necessary potential capability to match the level of complexity of the role. This is measured through assessing a person's perspective, which is a measure of the level of a person's information processing complexity or in other words his or her raw native ability,

    2. The skill, knowledge and expertise required to function in the role,

    3. Sufficient valuing of the work in the role, so as to be fully committed to doing the work.

Focus on Managerial Leadership

It is **not** part of any manager's leadership accountability to motivate or stimulate subordinates to do what they have been contracted to do. This means that managers can stop trying to change their behaviour so that it conforms to some idealised standard.

Leaders are leaders because they make leadership decisions based on short, mid and long-term planning. Great managerial leadership derives from the setting of sound strategies and success in carrying out those strategies and not from any particular personality qualities.

As such, CEOs can focus on developing sound strategic plans that will allow the achievement of the company's goals, set by the board or by the owner and, like all managers, they should seek the counsel of their subordinates in developing the strategy for successfully achieving these goals. Leading people means giving them assignments that contribute to success and displaying the initiative, flexibility, and adaptability to change direction and adjust assignments in order to overcome any difficulties along the way.

Some commonalities have been observed among leaders of the highest capability. Their self-awareness frequently includes an increasing sense of what they do not know, of what their experience does not teach them, of how much they still have to learn and, without any false modesty, how little they have yet achieved. The final sections of Winston Churchill's and Mandela's autobiographies, where one might expect a mature summary of all that has gone before, are entirely focused on the future, on how much there is still to be done.

'As for me, all I know is that I know nothing.' Socrates

J. Collins paper in Harvard Business Review, Jan 2001 regarding the key qualities of effective leaders in businesses identified that 'personal humility and professional will' are critical components. Collins recognised in his research that successful leaders are able to set their own needs aside.

Recognise That Recruitment is a Strategic Activity

Successful recruitment is all about answering three questions. Can the person do the job, does the person want to do the job and will they fit in around here?

Let's address the first one by taking the example of recruiting the next CEO of a large multinational financial services institution. It is likely that a Work Levels analysis of the role would show that the type of problems that the CEO would have to tackle:

- Involve a network of interconnected units with frequent and multiple linkages,
- Have arisen in the external environment of the unit,
- Are concerned with the long-term direction,
- Appear to have no clear boundaries or established context,
- Are presented as dilemmas between abstract ideas such as wealth creation and corporate social responsibility.

In order to undertake this role a person must be equipped with a range of skills, knowledge and experience. They must *also* have the required perspective (that is the mental model that they bring to bear in dealing with new and different situations) to enable them to get their head around the size, scope and complexity of the decisions that need to be taken. Even a lifetime in financial services does not necessarily equip everyone with this.

Exploring the context, expanding on the complexity of the decisions that need to be taken and defining the required perspective provides the basis for a well-constructed assessment process. When seeking to

139

understand what a person has actually achieved it is necessary to engage in a dialogue that is uncluttered by practised responses that so often occurs in interview situations. A rigorous approach is required and seemingly scientific psychological profiling falls short.

The outcome of a process that focuses on a person's technical skills, knowledge and expertise *and* their perspective will ensure that the appointed person will have what it takes to make decisions in conditions where precedent does not apply and there is a high level of ambient uncertainty.

That leaves us with the other recruitment-related questions, namely personal motivation and fit to the culture. Assessing motivation is easier than all those purveyors of psychometric assessment will have you believe. What is the person interested in, what draws their interest and attention? Ask them through a structured conversation and people will tell you.

If you have to ask 'will this person fit in around here?' it probably says more about your organization than it does about the person. Most people, if they are not suffering from acute depression or a whole host of other negative psychological conditions (e.g. paranoia) will tend to fit in unless you have a warped culture, such as one that emphasizes aggression or machismo behaviour.

Conclusion

You might consider the radical alternative to be idealistic. Guilty as charged! However, it is also achievable. It simplifies people and organizational management. For instance, it reduces the wasted amount of time that managers spend fretting about how to talk about performance issues. When commitment to a job is not a contested value, managers can focus on skills, knowledge and expertise. If a person, despite their commitment, cannot do a job, they need to be moved or fired after due process.

Chapter 20
Optimal Climates

In Chapter 2 I talked about how, after reading *Small is Beautiful,* I was left with a sense that management systems are great at establishing control but mostly stifle innovation and creativity. In this chapter, I provide a simple system that specifies the role of the line manager in achieving a balance between order and freedom.

It is the line manager who determines the working 'climate'. It is they who set the conditions for flexibility, creativity, the use of individual judgement and employee engagement. Sometimes this climate reflects the general climate of the organization and sometimes it is a direct result of the very specific style of the manager and the way that they empower, involve and motivate others.

In creating the climate, a manager must be able to balance, sometimes contradictory, behaviours. As this is a balancing act that is largely dictated by the needs of individual employees, there is no simple step-by-step competency formula.

Once management jargon and all HR ephemera are removed, managers can focus on the simple conditions that employees want and need.

These three conditions are:

- **Clarity;** people want to be clear about what they are expected to do,
- **Trust;** people want to feel that they can use their own judgement and be allowed to get on with work within their level of competence,
- **Purpose;** people want to know that the work undertaken is heading in the right direction and at the right pace and they want to know that the tasks undertaken link to a broader purpose.

These conditions are likely to be met if a manager demonstrates a balance of **Clarifying, Enabling** and **Reviewing**.

Clarifying consists of **Tasking and Involving**.

Tasking is about establishing the core purpose of a person's job and identifying their accountabilities. It is about specifying intended outcomes and a time-scale for completion of tasks. *Involving* is the process of ensuring that goal setting is not simply a one-way process. It requires recognition that the only real expert is the person doing the job.

Enabling consists of **Providing the Means and Space** to do the job required.

Providing the means involves equipping the person with the necessary tools, skills, knowledge and procedures that are safe and appropriate for the work to be undertaken. *Providing the space* for people is about creating the conditions for them to make their own judgements to the limit of their current capabilities.

Reviewing consists of maintaining **Contact** and making **Connection**.

Contact is monitoring without crowding. It ensures that the work assigned is still relevant to the organization and that the resources required are being used appropriately according to the current priorities. *Contact* also involves measuring performance against the intended outcome.

Connection is about recognising achievement. Most importantly of all, it is about communicating a sense of purpose and relevance for the work so that all individuals are clear as to how the job links to that of others and how the work is part of a broader goal.

The Optimal Climate

The optimal situation is where a manager provides a balance of Clarifying, Enabling and Reviewing. This leads to people making sensible decisions, which build good customer relations, cement supplier relationships and minimise rework and waste.

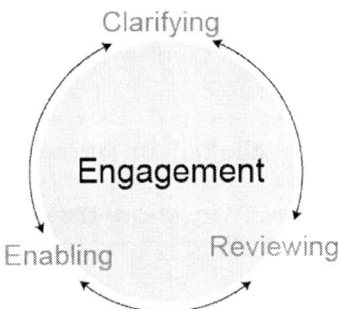

People are clear as to what is expected and what they hold accountability for. They feel that they are trusted to make their own judgements within their area of competence and they have a clear understanding as to how their work connects to a wider purpose. In short, they are engaged with their work.

Sub–Optimal Climates

As we know from our experience, not all managers operate in a balanced way. Some take on the organization style, which may stress one or more of these activities. Others just adopt what they have been shown or feel is the right style.

142

An imbalance of these management conditions has negative consequences and can lead to either 'foggy' or 'hot and barren' climates.

Foggy climates are established when managers, instead of Clarifying simply **Hand-Over** decisions to their subordinates. Instead of Enabling, they **Overly-Trust** a person to get on with their job in whatever ways that they feel are right. Instead of Reviewing, they **Neglect**.

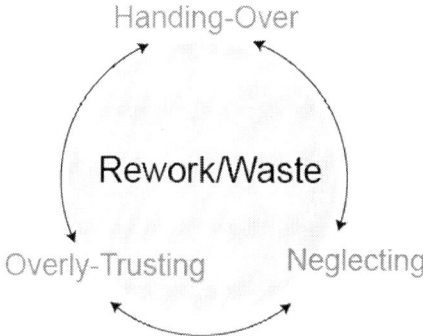

People can be given too much space to make their own judgements. The result is a diffuse environment in which people operate by guesswork. When there is guesswork, people generally feel more comfortable when they consult with everyone and when a collective view emerges. Therefore formal and informal meetings proliferate. A diffuse environment definitely results in rework and waste. Personal engagement is made difficult as ownership of outputs and accountabilities are unclear.

Hot and barren climates are established when instead of Clarifying, managers simply **Dictate** their requirements. Instead of Enabling, they **Disable** and see themselves as knowing best or they retain control over key parts of the role. Instead of Reviewing, they **Police** and recognise only non-compliance.

When people are given too little space and are driven by tasks and unrelenting processes, the result is a rigid working environment characterised by 'rule-following'. When there is rule- following, people carry on doing something that they know results in waste or poor quality. This jobsworth attitude is a classic symptom of disengagement. When senior people realise that things need to improve and introduce change initiatives, they wonder why these run into the ground.

360 degree feedback systems and employee-attitude surveys have become complete pains to managers and employees alike. Yet feedback is the lifeblood of learning.

Online survey techniques make it possible to measure the climate that is provided by managers. Collecting feedback from individuals across the organization informs managers about the quality of the 'space' that they provide to their subordinates and can also be aggregated to inform the Executive, 'what is it like to work for us - as an organization?' Now, that's 360 degree information that is worth the effort of compiling.

Chapter 21
Summary

You have accompanied me on a journey through a set of ideas that deal with time, complexity, uncertainty and the nature of *being*. In addition, I have presented a new model of human capability and a taxonomy that makes a dynamic link between the challenge of work and the capability required to undertake this. We have covered the implications of the new model of human capability within diverse fields. Now, as we draw to a close, it is time to introduce and identify some broad implications of all these ideas at a systems, leadership, organizational and individual level.

Systems

The future is not out there waiting to reveal itself. The future is imagined.

Given that we are all the architects of that future, is an overcrowded and polluted planet, managed by a super rich elite that values making money out of money more than anything else, the best system that we can imagine?

As I have attempted to make clear in earlier chapters, our behaviour is based on expectations and to a large extent the systems in operation set the expectations. However, economic doctrines or theories or systems never stand on their own feet; they are inevitably built upon our basic understanding of human nature, our outlook on life and its meaning and its purpose.

Systems are never more nor less than incarnations of human beings' most basic attitudes about themselves. E F Schumacher

Social and economic systems are based on a belief about human nature. If this nature is seen as essentially greedy the system exploits this aspect. If human nature is seen to be good we provide a pathway for leaders to flourish and wait for them to turn to the elements of truth and beauty. Where notions of human nature don't fit a prevailing view they are denied. For example, when human nature is seen to be complex and inconsistent this is denied by those pushing their version of 'rational man' and is replaced with a mistaken understanding that humankind has some sort of special status.

That is not to say that some systems are better than others. General evidence of material progress shows that free enterprise is the best instrument for the pursuit of personal enrichment. Our modern system in the form of liberal capitalism, at least in the UK and USA, ingeniously employs the human urges for personal gain as its motive power and attempts to overcome the most blatant deficiencies of *laissez-faire* capitalism by means of Keynesian economic management, a bit of redistributive taxation and (often toothless) regulatory bodies.

Can such a system conceivably deal with the problems we now face? These challenges include indebtedness, exhaustion of natural resources and the catastrophic collapse of biodiversity. The answer is self-evident; this system demands continuous and limitless economic growth of a material kind that cannot possibly fit, without proper regard for conservation, into a finite environment with finite resources. Conservation though comes a poor second despite all the fine words, world summits and Kyoto Agreements.

Materialism is the cornerstone value in free enterprise economies that have moved from producer to consumer societies. This is clear when we look at one of the main metrics our Government uses to measure economic success, namely Gross Domestic Product (GDP). It certainly can be argued that this was a false beacon of light, in a sea of metrics, which helped steer the economic ship on to the rocks.

GDP with its sense of production and added value sounds entirely positive. However, GDP also includes the retail industry. Retail is a wealth generator but it also is the vehicle for consumption. In the UK over 70% of GDP is connected to this sector. Up until the credit crunch, the GDP of most European nations increased at what seemed to be a healthy rate. However, this growth was, to a significant extent, a by-product of consumption based on easy credit and vast government borrowing. As spending in whatever form (earned or borrowed) is also included in GDP, it was no wonder that Europe looked like it had a sound economy.

If GDP stood for Gross Domestic Consumption it would not be used as the proxy indicator of a positive growing economy it has come to represent. In times to come, I suspect that GDP will be viewed as a self-deluded happiness index.

Certainly, humankind has built a system of production that ravishes nature and a type of society that promotes consumption. Money is considered to be all-powerful and the greatest skill of all, it seems, is to make money out of money.

The development of production capacity, consumption and the ability to make money from money are the highest goals of the modern world in relation to which all other goals, no matter how much lip service may be paid to them, take second place. This is leading us down a wretched road. Making money out of money has proved to be as beneficial to the whole of society as gambling is to the working person. Casino banking, just like all other gambling, makes huge rewards for those in control but leaves everyone else, on average, worse off.

146

Despite the upheavals, overall, our policy-makers cling to the illusion that we can be masters of our destiny. We are led to believe that the destructive forces of the modern world can be brought under control simply by mobilising more resources to fight pollution, preserve the ozone layer, discover new sources of energy and roll the money presses when things get tight.

We are turning to technology and money to cure the ills induced by our love affairs with technology and money. At the same time, the future that we imagine is becoming limited to shorter and shorter periods. Do we now have collective Attention Deficit Disorder?

Civilisation is revving itself into a pathologically short attention span. The trend is coming from the reducing time horizon of market economies, the distraction of multi-tasking and the ever-increasing overload of information. This shortening is evident in business. With its elevator pitches and tweets, it is the case that if you don't interest and entertain executives you won't rise above the noise, which they have surrounded themselves with.

Talking with teacher friends of mine, it seems that technology is hindering the development of attention spans. Pupils are so used to having their senses filled by the X-Box, PS3, online games, instant messaging etc., that it is increasingly hard to gain and hold their attention. Teachers have to be more like entertainers using technology such as interactive whiteboards and video-clips etc. When separated from the constant input and interaction through technology many young, and not so young, people become quite stressed.

Some sort of balancing corrective to the shortening of attention spans is needed. Where are the equivalents of the cathedral builders of days gone by who started projects that they knew would not be completed in their lifetime? If only we had a lens that encourages the long view and the taking of long-term responsibility, where long-term is measured in centuries.

Despite the precipice that the human herd is charging towards, will the leaders of China and the other rapidly developing nations alter course if they were to see where their dash for prosperity, and the need for more and more resources, is taking them and the rest of the world? Are they going to abandon the lure of materialism and instead first seek the kingdom of God? Unlikely, I know.

Even though unlikely, we need to start by imagining a different future. Two of the things that we must imagine and then bring into the present are 1) new ways of increasing the engagement of all citizens whilst providing clear boundaries and 2) an alternative system to that of the current form of liberal capitalism with its 'freeish' markets and 'weakish' regulatory systems, which better fits the new situation.

147

Leadership

Societies are made up of organizations and their leaders shape organizations.

We have just lived through a decade of greed. It had its icons, Steve Jobs and err... Steve Jobs spring to mind. It had its mission of making money from money and the gospel of shareholder value. It had its magic wand and the golden goose; the hedge funds and derivatives. Its dominant mindset was to achieve spectacular short-term results. Management and leadership often degenerated into sleights of hand or a follow-me at all costs mentality. The net effect was a culture that destroyed long-term value whilst it paid the management teams vast amounts.

The characters changed in the gold rush at the turn of the millennium but the story is essentially the same as in the greedy era of the 1980s. Therefore we cannot blame the current excesses on a single misguided generation or a unique set of historical circumstances.

The success at all costs myth has to change or the conditions for collective shortsightedness will spring up again. They surface whenever the conditions are right; for example when City folk harp on about the value of acquisition for its own sake, the media glorify glitz over substance and when political leaders equate every form of expansion with moral good and encourage destructive egoism in the name of healthy competition. J O'Neil

Since before the turn of the millennium, academics have been referring to new leadership models that address inherent uncertainty and recognise the need to treat organizations as complex adaptive systems. But the near collapse of financial systems, the rise of sovereign debt and dramatic fall of tax income has sent leaders back to old ways of thinking as they try to fix the plumbing. It can easily be taken that we have learnt nothing or certainly applied nothing of what we have learnt.

It seems that our actual leadership behaviours are unchanged and it might be a case of 'plus ça change, plus c'est la même chose', the more things change, the more they stay the same. But is there a good news story to tell? Yes there is, if we go back over centuries rather than focus on the last ten years. Despite the constant stream of grim news and shocking headlines, it is clear that humans are learning to appreciate each other more and leadership has profoundly changed as a consequence. This may not be apparent in close-up but with a wider lens it is clear that progress is being made due to the civilising influence of trade. When land-grabs and plunder were the main ways to access wealth, an appreciation of the other party might well have hindered progress. With the emergence of capitalism and the need to

148

base trade on contracts and trust, sensing and the capability to imagine the agenda of the other became a key 'trader' or 'commercial' skill. As Adam Smith pointed out, it is in our self-interest to know the other and the only way to do that is through imagination.

'Competition compels the wooer....to go out to the wooed, come closer to him, establish ties with him, find his strengths and weaknesses and adjust to them....competition achieves what usually only love can do; the divination of the innermost wishes of the other, even before he himself becomes aware of them'. George Simmel (1858-1918)

The necessity to appreciate the needs of a person is embedded in capitalism and as the speed of change increases there is a requirement to move beyond an understanding of needs and wants to embrace the complete the person. Leadership at the highest level stands on the cusp of a new era. Those who fully appreciate the challenge will sense and recognise emerging patterns and position him, or her self, personally and organizationally, as part of a larger generative force that will reshape the world. Leadership at the highest level is Level 7 Leadership and is about creating the value systems, social norms and boundaries by which the rest of us can live our lives. It is not about attaining a God Like presence.

Level 7 leaders understand that we live, lead and work in an era of dramatic upheaval. The waves of change sweeping the world including digitalisation, globalisation, demographic shifts, migration and individualisation as well as the rapid degradation of social and natural capital are giving rise to clashing forces. These play out as tensions between multiple polarities: speed and sustainability, exploration and exploitation, global and local ways of organizing and order and freedom.

Although general statements about the increasing change of pace have been true at many times and places in human history, there is something different about today's circumstances. The pace of change is faster. As change begets change, the frequency and amplitude of restructuring and reforming are significantly greater and the pathways of emerging futures are less predictable than they were in earlier times.

As the economic foundations of our business world are transformed from stable to more dynamic patterns the nature of leadership has to change too. For Level 7 leaders what is 'real' is changing. In traditional and more stable business environments, mental and social processes were considered peripheral complications in a value chain largely based on the primacy of the tangible world.

In order to succeed, Level 7 leaders will have to develop a new cognitive capacity that involves paying attention to the intangible sources of knowledge and knowing. Thus, the challenge is to develop higher qualities of pattern recognition by shifting the emphasis from understanding a system in operation to becoming more mindful of the deep sources from which behaviour and profound innovation and change emanate.

A critical role of the Level 7 leader is setting the conditions for others and providing a context for them to work within. Context setting is about recognising that people are meaning making and meaning seeking beings. The process of creating meaning involves the telling of stories. Stories capture exactly those elements that formal decision-methods leave out. Logic tries to generalise and reduce arguments to facts that are removed from emotions. Stories capture the context and the emotion. Increasingly, as people have more choice as to where they live and work, they are developing a hunger for what stories can provide; a deeper understanding of how we fit in and why that matters. The requirement of leaders to encapsulate, contextualise and emotionalise has become vastly more important as the waves of change that people are subjected to increase in frequency and amplitude.

Human activity arises out of locomoting towards an intended future. How we imagine the future influences our behaviour. When Level 7 leaders fully appreciate that people are locomoting toward an imagined future, it is a wake up call to the architects of corporate and social values to create an imagined future that is worth locomoting towards.

Organizations
The history of the industrialisation of Britain throws up many examples of entrepreneurs and business owners who saw each person as an irreducible mystery and not just as a pair of hands to be used and then discarded. These enlightened employers regarded their workers as more than necessary costs. This manifested itself in paternalist approaches, which often extended to building model villages such as at Bourneville and Port Sunlight.

Despite its best intentions, paternalism has waned. The progressive strengthening in the free market economies of the idea that the individual is the agent in charge of their own life and responsible for their own choices has made paternalism old fashioned. Now, more sophisticated methods are being looked for as employers seek to engage their workforce.

It could be argued that engagement is more important now because organizations in all sectors need the imagination and discretion of their employees. Organizations need people to make decisions in the face of uncertainty and to use their discretion as they make products or serve customers and clients. Organizations also need people to use their imagination to create new ways of controlling costs and limiting waste, to develop new products and new markets, to find better ways of serving customers and to see opportunity in risk. In a globalised world where patterns of influence are shifting, people need imagination to understand and collaborate with partners from diverse backgrounds who have different ways of making sense of the world.

Despite their best wishes, few organizations are characterised as being full of engaged people. For those organizations not in such a fortunate position, I suggest that they start by questioning the beliefs that underpin the employment relationship.

Webster's Dictionary defines Work as 'labour, travail, toil, drudgery, grind'. Based on such a definition, many believe that people work only because they have to. It is also believed that the nearer that you are to the top of the employment hierarchy the more satisfaction you gain from being employed. By the same token, it is assumed that the lower you are, the more frustration and dissatisfaction you experience with work. People at lower levels are viewed as needing constant supervision. In addition, it is believed that everyone needs results-based financial incentives to spur them to produce high quality work. This attitude implies that people are paid basic wages/salaries for being less than fully committed to performing their work satisfactorily. The above are limiting beliefs and are the basis for employment practices that revolve around control and coercion.

Employee engagement is dependent upon treating people at all levels as people who can make independent choices.

In treating people as people, organizations will recognise that managing people is messy; fortunately they don't conform to the same laws as machines and as such creativity and judgement can lead to unexpected and wonderful outcomes. They will see people not as objects or resources that can be switched on and off like electrical apparatus. Instead they will understand that motivation is generated from within the person. Motivation cannot be imposed; it is a choice that individuals make. They will respect the individual and truly believe that all individuals are purposeful, intentional creatures. They will understand that the choices we generate are based on the meaning that we make of our situation.

When a manager absorbs the general attitude that people are irreducible mysteries and provides the optimal climate, people give more of themselves as they are treated as human beings who can make choices.

Individual Level

Change starts at home.

'God grant me the serenity to accept the things that I cannot change,

courage to change the things that I can and the wisdom to know the difference.' Reinhold Niebuhr

In developing the wisdom to know the difference between what we can change and what we can't, we need to understand that we are not isolated agents impacting through the force of our will. We are codependent and contingent.

As social creatures we are continually in the process of relating to others. Relating is, to a large extent, about sharing space. It is regulated by the distance and proximity of others and plays itself out as a contest over territory of various forms. In the constant battle to assert our rights and exercise power it is easy to miss that we are formed through the 'other' that we encounter. The way in which we relate to others constitutes one aspect of what we become ourselves. For instance, in the submission of others we become a dominator.

Change is possible when we develop a mature relationship with time and uncertainty. Through this it is possible to understand what we can and cannot control.

Each situation is a new combination of elements, like waves in an ocean rolling together and forming new patterns. We are neither at the mercy of these waves nor are we able to choose our way across deliberately and singlehandedly. Rather, we are capable of learning to ride the waves and occasionally plunging into the surf. The total outcome is unpredictable and we can never retrace our steps. Emmy Van Deurzen

We build for the future when we accept that there is much that we can't predict and control whilst still being prepared to seize the initiative. Also, we make the future clearer when we recognise the fundamentals that won't change. These are the timeless principles of human and social interaction and life success that endure. Instead of asking what is going to change, it helps to ask, what's not going to change?

'I live in the past, present and the future. The spirits of all three shall strive within me. I shall not shut out the lessons that they teach.' Charles Dickens

It is helpful to keep track of our time horizons. Where are we spending most of our thinking time; in the past, present or future?

'In this great future you can't forget your past.' Bob Marley

Our past provides the clearest indicator of what we will enjoy in the future.

We must live in the present whilst drawing from the experiences of the past and enjoying the flow of our current life. But we also need to prepare for the future. It is the courage to look ahead at future challenges and put in place grounded plans that will help us get more of what we have enjoyed in the past.

It is easy to allow our past to become our future in which we rewind past experience to relive previous hurts or grievances. We should appreciate what is good about our upbringing, our early life experience, school years, relationships, bosses and so on. And if it wasn't that good, we should put it behind us, imagine a different future, and move on. After all, we are the stories that we tell ourselves.

We are our stories. We compress years of experience into a few compact narratives that we convey to others and tell ourselves. Daniel H Pink

To really be the architect of our own lives we need to recognise the power of language and not get hypnotized by our own inner voice. When we hear:

- 'I can't move on'......it means, 'I choose not to address the issues about moving on',
- 'I don't know what I want'..... it means, 'I know what I want. I want a number of things but I am unwilling to choose just one of them',
- 'I can't decide'...... it means, 'I cannot make a decision because any decision I do make will have downsides and I am not willing to accept that',
- I have no ideas as to what is going on'..... it means, 'I know exactly what is going on but the idea of doing anything about it is threatening and I would prefer not to think about it'.

It's also easy to get caught up in the priorities and pressures of the present, becoming so busy we forget why we're busy. We have to have a mature outlook on time that helps us walk a tightrope, balancing two very different tendencies. On the one hand, fear of what lies ahead can make us too timid. On the other hand, fantasy or hubris can make us bound on with unfounded confidence.

Instead of trying to avoid uncertainty, we need to first recognise that this is the very essence of life. There is a huge literature out there on worry and how to stop troublesome thoughts. What I have learnt (yet often fail to practise) is that the first step is to recognise that worry and anxiety are part of the human condition.

Worry never robs tomorrow of its sorrow; it only saps today of its strength. A J Cronin

Shakespeare knew a thing or two about human nature.

'Things won are done; joy's soul lies in the doing' William Shakespeare

In understanding that the journey is more important than the destination Shakespeare would also appreciate that how we undertake the journey shapes us. To everything there is a consequence and the ends rarely justify the means. The route that we choose to attain our goals will change us and not always in ways that are conducive to our long-term happiness. If, having achieved our original goal, we're wondering why we are still not happy it is likely that the means that we selected to achieve our goals turned us into a different person.

Although we inhabit a world that mostly values the continual acquisition of material things and experiences that come from expensive and environmentally damaging travel, existential questions enter on occasion unbidden and sometimes unwelcome. These existential questions are beguilingly simple; who am *I*, what am *I* doing here and where am *I* going? These are the questions that move us into the spiritual dimension.

The spiritual dimension begins early in life. Young children ask profound 'why' questions especially when they realise that there is such a thing as death. We know we will die and the prospect creates anxieties. The fear of death is of course a healthy instinct; it's hard wired in us to ensure our evolutionary survival.

The spiritual world is the meta-world where all the rest of our experience is put into context. It is the world of ideas, faith, belief and meaning. In this dimension of our existence we really come into the true complexity of being human as we organize and transcend other levels of existence and create a philosophy of life, defining our personal stance and world-view. Of course, not everything has to be about God or the Holy Spirit; it is possible for people to raise specific ethical or political systems to the status of an over-arching belief system.

People cannot live without beliefs and values. Immersed in a shopping culture we are encouraged to believe that things and experiences will make us happy. Technology, advertising and the ability to get instant answers through the Internet means that we know how to do a lot of things. It also means that we often possess a lot of things. Whilst we know and have a lot, do we know what to do?

The spiritual dimension sounds like it is at a higher level than the physical, social or psychological dimensions but it is constituted by our most basic convictions; by those ideas that really have the power

154

to move us. In other words, this dimension consists of metaphysics, ethics and ideas that, whether we like it or not, transcend the world of facts. They cannot be proved or disproved by ordinary scientific method but that does not mean that they are purely subjective or relative or secondary to buying and selling. However, the sad reality is that through taking on prevailing ideas and received wisdom, we have degraded words such as virtue, love and temperance. As a result we are uneducated in the subjects of ethics.

The spiritual dimension is where we have to create an orderly system of ideas about ourselves and the world that we inhabit. The output provides the direction for our various strivings. If we have been too busy in the day-to-day hurly-burly to pay this sufficient attention the spiritual dimension will not be empty; it will be filled with a set of dominant ideas and received wisdom which, one way or another, have seeped into our mind without proper consideration

Given the shopping culture and the prevailing ideas of life, and its futility, that are around today the thoughts that seep into our heads are likely to be a total denial of meaning and purpose of human existence on earth. Fortunately, the human heart is more intelligent than the head and refuses to accept these ideas in their full weight. We are saved from despair by the power of the human heart but nevertheless left in a state of confusion.

In confusion, what do we turn to now that we have abandoned our classical Christian heritage?

In moments of doubt and confusion we turn to psychotherapy and counselling. However, these disciplines largely reinforce a belief that we *are* our own masters. Yet, there are many signs that our trust in our autonomy may be misplaced. With an overpopulated and polluted planet, we need to take a close look at the processes in which we take part and to which we owe our living.

Everywhere people ask, 'what can I actually do?' The answer is as simple as it is disconcerting; we can each of us work to put our own inner house in order. The guidance we need for this work cannot be found in science or technology but it can still be found in the traditional wisdom of mankind. E F Schumacher

On Human Nature

I trust that, if you have followed me so far, you have enjoyed the views along the way and had glimpses of the peaks to which I referred in the introduction. However, didn't I also promise a new perspective on human nature? I hope that you won't be disappointed with my conclusion.

Humans have a range of emotions and demonstrate a remarkably wide range of behaviours but when we try to summarise 'human

155

nature' we produce inaccurate stereotypes and limiting beliefs. When these stereotypes and limiting beliefs are the basis for our economic, social and organizational systems the result is inevitably a self-fulfilling cycle. Systems are established to exploit the desire for personal gain and, lo-and-behold, people seem to be greedy. Capitalism is based on a conception of human nature and so too is Marxism.

Human nature is as complex and diverse as anything in nature and caricatures don't do it justice. There is the capacity for harmony and the utmost cruelty, greed and altruism and, what is more, each individual has these capacities. The truth about life and human nature is that it is more complex and diverse than we can imagine.

If human nature can be reduced to a single characteristic it is that we have a deep-seated need to avoid the anxiety of uncertainty. Uncertainty, what uncertainty? Whenever we can, we eradicate it. We seek clarity and have the mental software that enables a range of predictions to be made. This power of prediction (not necessarily accurate) gives us a sense of control and our anxiety is reduced even if this means that we have to tell ourselves stories that mask the fact that we jumped to conclusions.

We might try to avoid it but uncertainty about the future is the locus of our freedom. Once we realise that the future is uncertain and waiting to be created by the way we imagine, will, choose and accept it to be, we are liberated from the implicit determinism of perfect rationality and biological mechanisms. Uncertainty provides the genuine existential freedom that is the gift of life.

Postscript

For a while I bought into the power of classic success thinking. When it comes to goal setting, this assumes our current-self knows what our future-self wants and what will make us happy. It emphasises 'big dreams and practical goals'. Think big thoughts about what is possible; create a vision of what that looks like and then set specific goals to achieve this vision. But the success formula hasn't worked for me. More specifically it hasn't worked for me in the economic sense of delivering financial security.

The problem is that the ideas on which I have tried to shape a business were outside of a prevailing paradigm. In taking onboard the ideas of outsiders, and developing these further, I have, as a result, unwittingly become an outsider. I now know that outsiders are largely ignored. It is like donning a coat of invisibility. In looking for a 'bright side' I suppose I can be pleased that I am not moving in academic circles. As such, I have simply been invisible rather than the subject of withering invective that academics have developed into a fine art.

The goal of financial security has not been met but if success is recast to mean the quality of relationships and the opportunity to get to know remarkable people, my life has been rich indeed.

Whilst I recognise that my understanding of life is still developing, I hope that I have a mature stance to the future incorporating an appreciation of time, complexity, uncertainty and the nature of *being*. Certainly, I now know what is within my power to change. As such, the strategies and tactics that I deploy will be based more on flexible opportunism than on any predetermined blueprint that I have developed for my future life. I accept that life can't be lived like a project plan. Instead I am keeping an open mind, alert to what is and what isn't working and recognising that my values and priorities will change.

If you are interested, watch this space! Let's see if purposeful drifting provides a better alternative to the process of meaningful living than all the 'success recipes' that are thrown our way in contemporary pop/business psychology.

Life is a journey and writing this book has been a journey too. In all journeys the unexpected happens. In this case, I wasn't fully prepared for the realisation that when you bring time, complexity, uncertainty and the nature of *being* centre-stage everything changes.

Talking of journeys, I would like to end with a verse from the poem, Ithaka.

Keep Ithaka always in your mind.
Arriving there is what you're destined for.

157

But don't hurry the journey at all.
Better if it lasts for years
so you're old by the time you reach the island
wealthy with all you've gained on the way,
not expecting Ithaka to make you rich.

Constantine Cavafy

Glossary of Terms

Accountability: A situation where an individual can be called to account for his or her actions by another individual or body authorised both to do so and to give recognition to the individual for these actions.

Awareness: The primitive or basic feature of living organisms whereby they take note of either external or internal worlds by what ever sensory means that they have available. Awareness does not imply the ability to describe in words what the organism is aware of.

Being: The life force that provides for individual agency and the creation of intentions and free will.

Capability: The ability of a person to do work. **Applicable Capability**: The capability that someone has to work in a specific role at a given level at the present time. **Current Potential Capability**: A person's highest current level of work complexity. It is the maximum level at which someone could work at the present time, provided that the work is perceived to be of value and given the opportunity to acquire the necessary skills, knowledge and expertise. **Future Potential Capability**: The maximum level at which a person will be able capable of working in the future.

Choice (and Choosing): That part of the process of work in which a possible action (a choice) or possible actions (choices) are garnered, one of which might be decided (acted) upon.

Complexity: Complexity is determined by the scale and scope of the decisions to be made, the resources available to make the decisions and the time-scale over which the impact of the decisions play-out.

Conscious(ness): Linguistically articulated awareness including 'self-talk'.

Decision: The taking of actions with the commitment of resources that ends a process of mulling with its judging and choosing of possibilities.

Disengagement: A process only open to humans characterised by withdrawal from engaged and focused external attention on the here and now, to attention focused on the internal world of language.

Dynamics (behavioural): The study of the movement of organisms engaged in goal directed behaviour.

Future: A current state of attention comprising things an organism might like to do or say, or anticipates doing or saying. The future exists in the present; we are neither moving towards the future nor is it coming towards us.

Ineffable: Not accessible to being expressed using words.

Information: Any kind of data or knowledge. **Information complexity**: The number of factors, the rate of change of these factors and the ease of identifying the factors in a situation. **Orders of Information Complexity**: A series of step changes in the complexity of information processed.

Intention: A decision by an organism to attempt to satisfy a need.

Knowledge: Awareness of how to do something.

Locomoting: The movement of any organism towards its goal

Ontological: The features of behaviour and development that are specific to each individual and thus vary from individual to individual.

Organical: Of or relating to living systems.

Perspective: This is a person's 'focal width' and describes the breadth and depth of their mental model, which is utilised to make sense of their world.

Phylogeny: The genetically determined features of behaviour and development that are common to some or all individual members of a species.

Physiology: The study of subsystems of living organisms.

Systems Theory: An explanation of a system in terms of the interaction of its parts. Open system: A system that is unbounded, and because it is unbounded, its precise conditions will never repeat again.

Time: An ordering of experience by living organisms in terms of events occurring at earlier and later movements in relation to each other.

Kairos: The time axis of intention, organized in terms of past, present and future, all in the present. **Chronos**: The time axis of occurrence or of clocks used to measure how long an event actually took from beginning to end. **Time-Horizon**: The method quantifying an individual's potential capability, in terms of the longest time-span that they can handle. **Time-span of Discretion**: The target completion of the longest task or task sequence in a role. Time-span measures the Work Level of the role.

Trust: The ability to rely on others to be truthful and to behave in a helpful and understanding way.

Work: The exercise of judgement and discretion in making decisions in carrying out goal-directed activities. **Work Level**: The weight of responsibility felt in roles as a result of the complexity of the work in the role. The Work Level in any role can be measured by the time-span of discretion of the tasks in that role.

References

Ariely D. (2008) *Predictably Irrational.* Harper Collins

Bannister D and Fransella F. (1971) *Inquiring Man.* Penguin

Bergson H. (1911) *Matter and Memory.* Reprint, Zone Books 1988

Bodmer W and Cavalli Sforza L (1970) *Intelligence and Race.* Scientific American October

Bones C. (2007) *Engagement is at the heart of successful M&A.* Ivey Business Journal

Boudreau J, and Ramstad P. (2007) *Beyond HR; The New Science of Human Capital.* Harvard Business School Publishing Corporation

Bovens M. (1998) *The Quest for Responsibility, Accountability and Citizenship in Complex Organizations.* Cambridge University Press

Boole G. (1859) *Investigations of the Laws of Thought.* Reprint, Dover Publishing

Brown D and Dive B. (2009) *Level Pegging.* People Management

Charam R, Drotter S and Noel J. (2011) *The Leadership Pipeline: How to Build the Leadership Powered Company.* Wiley

Collins J. (2001) *Good to Great.* Random House

Connor R and Mackenzie-Smith P. (2002) *The Leadership Jigsaw – finding the missing piece.* Business Strategy Review, Volume 14 Issue 1, pp 59-66

Connor R. (2009) *Responsible Governance and Human Resource Management: the connection.* Developing HR Strategy, Issue number 27, Wolters Kluwer (Ltd) Kingston

Connor R. (2009) *The Credit Crunch- You get what you pay for and other important lessons*, Developing HR Strategy, Issue number 28, Wolters Kluwer (Ltd) Kingston

Connor R. (2011) *The Ten Commandments That Build Healthy, Successful and Long Lasting Organizations.* Ecademy Press

Connor R. (2012) *It's About Time.* emp3books

Connor and McGuire R. (2010) *Job Description and Person Specification Alignment- the unlikely hero in the search for high performance*, Developing HR Strategy, Issue number 30, Wolters Kluwer (Ltd) Kingston

Csikszemtmihalyi M. (1988) *Optimal Experience.* Cambridge University Press.

Dawkins R. (2009) *The Greatest Show on Earth.* Bantam Press

Dennett D. (1996) *Elbow Room.* Cambridge MIT Press

Dive B. (2002) *The Healthy Organization.* Kogan Page, London

Dive B. (2008) *The Accountable Leader.* Kogan Page, London

Ehrenreich B. (2010) *Smile or Die.* Granta Publications

Friedman M. (1982) *Capitalism and Freedom.* University of Chicago

Press

Galbraith J K. (1987) *A History of Economics*. New York, Penguin Books

Gladwell M (2005) *Blink*. Penguin

Gladwell M (2008) *Outliers*. Penguin

Hancock and Zahawi (2012) *Masters of Nothing*. Biteback Publications

Heidegger M (1962) *Being and Time*. Blackwell

Hills J, Sefton T and Stewart K (eds). *Towards a More Equal Society?* Policy Press p 26-27

Jaques E. (1952) *The Changing Culture of a Factory*. Reprint, Gloucester: Cason Hall

Jaques E. (1964) *Time-span Handbook*. Reprint, Gloucester: Cason Hall

Jaques E. (1976) *A General Theory of Bureaucracy*. Reprint, London Gregg Revivals, 1993

Jaques E. (1982) *Free Enterprise, Fair Employment*. Reprint, Gloucester: Cason Hall

Jaques E. (1988) *Creativity and Work*. Madison, International Universities Press

Jaques E. (1990) *The Conscious, Preconscious and Unconscious Experience Called Time*. Reprinted in Creativity at Work. Madison, International Universities Press

Jaques E. (1990) *Time and Measurement of Human Attributes*. Reprint, Creativity and Work, Madison, International Universities Press

Jaques E. (1992) *Why the Psychoanalytical Approach to Understanding Organizations is Dysfunctional*. Human Relations, Vol 48

Jaques E. (1998) *Requisite Organization*. Cason Hall

Jaques E. (2002) *The Life and Behaviour of Living Organisms*. Praeger

Jaques E. (2002) *Social Power and the CEO*. Quorum Books

Jaques E and Cason K. (1994) *Human Capability*. Cason Hall

Jaques E and Clement S. (1991) *Executive Leadership*. Gloucester, Cason Hall and Oxford, Basil Blackwell

Jaques E, Gibson, R and Isaac J. (1978) *Levels of Abstraction in Logic and Human Action*. Reprint, Gloucester, Cason Hall

Jaques E and Zinke W. (2002) *The New Adult Stage*. Boulder, Human Resources Services

Kahneman D. (2011) *Thinking, Fast and Slow*. Penguin

Kelly G. (1955) *A Theory of Personality*. Penguin

Knight F. *(1921) Risk, Uncertainty, and Profit*. Boston, MA: Hart, Schaffner & Marx; Houghton Mifflin Company

Lansley S. (2012) *The Cost of Inequality*. Gibson Square

Lewin K. (1935) *A Dynamic Theory of Personality*. New York, McGraw-Hill

Lewin K. (1936) *Principles of Topological Psychology*. New York, McGraw-Hill

Lewin K. (1952) *Field Theory in Social Science.* London, Tavistock Publications

Levitt S and Dubner S (2005) *Freakonomics.* Penguin

Lindley D. (2006) *Understanding Uncertainty.* Sage Publications

Maturana, Humberto and Varela. (1986) *The Tree of Knowledge.* Boston Shambhala

Michaels, Handfield-Jones & Axelrod. *War for Talent.* Harvard Business School Press

Munro A. (2011) *Now It's About Time.* Amazon Distribution

O'Neil J. (1993) *The Paradox of Success.* McGraw Hill

Pink D. (2006) *A Whole New Mind.* Penguin

Pinker S. (1994) *The Language Instinct.* Penguin

Popper K. (1963) *Of Clouds and Clocks. In Objective Knowledge.* Oxford, Clarendon Press

Raynor M (2007) *The Strategy Paradox.* Doubleday

Richardson R. (1971) *Fair Pay and Work.* London; Heinemann

Robson C. (2010) *Confessions of an Entrepreneur.* Pearson Education

Rowbotton R and Billis D (1987), *Organization Design: The Work Levels Approach.* Gower, London

Senge P. (1990) *The Fifth Discipline.* New York, Doubleday

Shackle G. (1972) *Epistemics and Economics.* Cambridge University Press

Shepard, K (Editor) (2007), *Organization Design, Levels of Work and Human Capability.* Global Organization Design Society

Schumacher (1973) *Small is Beautiful.* Blond and Briggs

Stamp G. (1978) *Assessment of Individual Capacity.* In Jaques et al Levels of Abstraction in Logic and Human Action. Reprint Gloucester, Cason Hall

St Augustine. (1960) *Confessions.* New York, Image books

Stone B. (1997) *Confronting Company Politics.* Macmillan Business

Toffler A. (1970) *Future Shock.* Bodley Head

Van Clieaf M. (2006) *PEMs: The Magic Bullet?* Executive Compensation Strategies Volume 2 Number 4

Van Clieaf M and Langford Kelly J. *The New DNA of Corporate Governance: Strategic Pay for Strategic Value.* Corporate Governance Advisor, Volume 13 Number 3

Van Deurzen. (1997) *Everyday Mysteries.* Routledge

Varela F, Thompson E and Rosch E. (1996) *The Embodied Mind.* Cambridge. MIT Press

VonBertalanffy L. (1968) *General Systems Theory.* New York, Braziller

Biography of the Author

Russell has over 30 years' experience of working and consulting for both national and multi-national organizations.

He has been influenced by the ideas that come under the broad umbrella of 'Socio-technical Systems' and particularly by the work of Elliott Jaques. It was at the London School of Economics that Russell was first introduced to the pioneering thoughts of Elliott Jaques and, many years later, was reintroduced to his work through an association with the Brunel Institute of Organization and Social Science. Elliott Jaques' work provides the foundation for a total system of effective organization and managerial leadership in the 21st century and his ideas have been highly influential in the design and development of a number of top organizations.

Russell has developed The Work Levels Taxonomy™. This is a comprehensive classification and measurement system that segments job contribution and people capability into distinct levels and includes best-practice skills, knowledge and expertise frameworks. It is a new system based on the fundamental principles established by Jaques and his followers over the last forty years. It is possible to find out more about the taxonomy by visiting www.7-levels.com.

Russell is the author of two other books, *It's About Time* and *The Ten Commandments That Build Healthy, Successful and Long Lasting Organizations*.

Lightning Source UK Ltd.
Milton Keynes UK
UKOW030025300413

209959UK00009B/230/P